Complement Methods and Protocols

METHODS IN MOLECULAR BIOLOGY™

John M. Walker, SERIES EDITOR

METHODS IN MOLECULAR BIOLOGY™

Complement
Methods and
Protocols

Edited by

B. Paul Morgan

Department of Medical Biochemistry
University of Wales College of Medicine
Cardiff, UK

Humana Press ✳ **Totowa, New Jersey**

© 2000 Humana Press Inc.
999 Riverview Drive, Suite 208
Totowa, New Jersey 07512

This publication is printed on acid-free paper. ∞

ANSI Z39.48-1984 (American Standards Institute) Permanence of Paper for Printed Library Materials.

Cover design by Patricia F. Cleary.

Cover illustration:

For additional copies, pricing for bulk purchases, and/or information about other Humana titles, contact Humana at the above address or at any of the following numbers: Tel: 973-256-1699; Fax: 973-256-8341; E-mail: humana@humanapr.com, or visit our Website at www.humanapress.com

Printed in the United States of America. 10 9 8 7 6 5 4 3 2 1

Library of Congress Cataloging-in-Publication Data

Complement methods and protocols / edited by B. Paul Morgan
      p. cm. -- (Methods in molecular biology; v. 150)
   Includes bibliographical references and index.
   ISBN 0-89603-654-5 (alk. paper)
   1. Complement (Immunology)--Laboratory manuals. I. Morgan, B. Paul. II. Series

  QR185.8.C6 C685 2000
  616.07'997--dc21

                                 99-058849

Preface

The complement system, first described more than a century ago, was for many years the ugly duckling of the immunology world, but no more. Complement in recent years has blossomed into a fascinating and fast moving field of immediate relevance to clinical scientists in fields as diverse as transplantation biology, virology, and inflammation. Despite its emergence from the shadows, complement retains an unwarranted reputation for being "difficult." This impression derives in large part from the superficially complicated nomenclature, a relic of the long and tortuous process of unraveling the system, of naming components in order of discovery rather than in a systematic manner. Once the barrier of nomenclature has been surmounted, then the true simplicity of the system becomes apparent.

Complement comprises an activation system and a cytolytic system. The former has diverged to focus on complement to distinct targets—bacteria, immune complexes, and others—so that texts now describe three activation pathways, closely related to one another, but each with some unique features. The cytolytic pathway is the same regardless of the activation process and kills cells by creating pores in the membrane. Complement plays an important role in killing bacteria and is essential for the proper handling of immune complexes. Problems occur when complement is activated in an inappropriate manner—the potent inflammation-inducing products of the cascade then cause unwanted tissue damage and destruction.

Complement's renaissance has been driven in large part by the discovery of the complement regulatory molecules and the realization that these molecules and other agents can provide effective anticomplement agents for use in therapy. As newer and better anticomplement agents become available, the requirement for laboratories to assess complement activation in clinical samples and to monitor the effects of anticomplement agents will grow.

Complement Methods and Protocols aims to provide a comprehensive source of up-to-date protocols for the study of the complement system, both for the basic scientist interested in understanding the mechanisms of activation and the clinical scientist wishing to quantify complement activation. In the first chapter, the complement system is briefly reviewed to set the stage for the methods chapters to follow. The next two chapters describe methods for purifying complement components, using classical chromatography and immunoaffinity approaches, respectively. Chapters 4 to 6 describe methods for the functional analysis of complement components, regulators, enzymes, and complexes, including a detailed description of the generation of the depleted sera essential for complement assays. Methods for measurement of complement activation fragments and complexes deposited on cells, in tissues, or in biological fluids are detailed in Chapters 7 to 10. Chapter 11 provides an overview of screening methods for identifying and assessing complement deficiency and Chapter 12 a detailed account of methods needed to assess deficiency of C1 inhibitor. Other clinically relevant protocols for analysis of complement autoantibodies, immune complexes, and complement allotypes are provided in Chapters 13 to 15. Chapter 16 departs from the main theme of the book to describe protocols for generating gene-deleted mice, included here because of the enormous influence such methods are now having on complement research. The final chapter reviews complement deficiencies in experimental animals, listing the different complement deficiencies defined in animals and the experimental models in which these deficient animals have been examined.

I am grateful to my friends and colleagues who have contributed to this volume for their willingness to make time in their busy schedules. In particular, I wish to thank the members of the Complement Biology Group in Cardiff, many of whom have contributed chapters to this volume and others who have reviewed parts of the manuscript or contributed to the tedious task of assembling the appendices. I promise I won't do it again in a while! Finally, thanks to The Wellcome Trust for their continued and generous support of complement research in Cardiff.

B. Paul Morgan

Contents

Contributors

MARINA BOTTO • *Rheumatology Section, Division of Medicine, Imperial College School of Medicine, Hammersmith Campus, London, UK*

ANNE E. BYGRAVE • *Rheumatology Section, Division of Medicine, Imperial College School of Medicine, Hammersmith Campus, London, UK*

KEVIN A. DAVIES • *Rheumatology Section, Division of Medicine, Imperial College School of Medicine, Hammersmith Campus, London, UK*

C. ERIK HACK • *CLB and Department of Internal Medicine, Academic Hospital of the Free University Amsterdam, Amsterdam, The Netherlands*

JUHA HAKULINEN • *Department of Bacteriology and Immunology, Haartman Institute, University of Helsinki, Helsinki, Finland*

CLAIRE L. HARRIS • *Department of Medical Biochemistry, University of Wales College of Medicine, Heath Park, Cardiff, UK*

STUART LINTON • *Department of Medical Biochemistry, University of Wales College of Medicine, Cardiff, UK*

SEPPO MERI • *Department of Bacteriology and Immunology, Haartman Institute, University of Helsinki, Helsinki, Finland*

B. PAUL MORGAN • *Department of Medical Biochemistry, University of Wales College of Medicine, Heath Park, Cardiff*

JULIAN T. NASH • *Rheumatology Section, Division of Medicine, Imperial College School of Medicine, Hammersmith Campus, London, UK*

PETER NORSWORTHY • *Rheumatology Section, Division of Medicine, Imperial College School of Medicine, Hammersmith Campus, London, UK*

ANN ORREN • *Department of Microbiology, National University of Ireland, Galway, Galway, Ireland*

O. BRAD SPILLER • *Department of Medical Biochemistry, University of Wales College of Medicine, Heath Park, Cardiff, UK*

ANTTI VÄKEVÄ • *Department of Bacteriology and Immunology, Haartman Institute, University of Helsinki, Helsinki, Finland*

CARMEN W. VAN DEN BERG • *Department of Pharmacology, University of Wales College of Medicine, Heath Park, Cardiff, UK*

REINHARD WÜRZNER • *Institut für Hygiene, Innsbruck University, Innsbruck, Austria*

1

The Complement System: *An Overview*

B. Paul Morgan

1. Introduction

The complement (C) system consists of a group of 12 soluble plasma proteins that interact with one another in two distinct enzymatic activation cascades (the classical and alternative pathways) and in the nonenzymatic assembly of a cytolytic complex (the membrane attack pathway) (**Fig. 1**; **Table 1**). A third activation pathway, termed the lectin pathway, has recently been described *(1,2)*. Control of these enzymatic cascades, essential to prevent rapid consumption of C in vivo, is provided by 10 or more plasma and membrane-bound inhibitory proteins acting at multiple stages of the system. C plays a central role in innate immune defense, which provides a system for the rapid destruction of a wide range of invading microorganisms.

The purpose of this volume is to provide a balanced account of the methods which have been applied to the study of the C system in clinical and research laboratories. In the early chapters, methods for the isolation of the individual C components, regulators and receptors will be described and assays for the measurement of C activity in the various pathways will be detailed. Later chapters will describe aspects of C methodology of relevance to the clinical laboratory—protocols for screening for C deficiency and deficiency of C1 inhibitor, methods for measurement of C activation products in biological fluids and tissues and methods for allotyping C components. Other chapters will describe advanced technologies which are now having enormous influence on C research—structural analysis of the components and regulators and gene targeting to generate animals deficient in individual components and regulators.

In order to appreciate the methodologies to be described, it is essential first to understand the basics of the system.

From: *Methods in Molecular Biology, vol. 150: Complement Methods and Protocols*
Edited by: B. P. Morgan © Humana Press Inc., Totowa, NJ

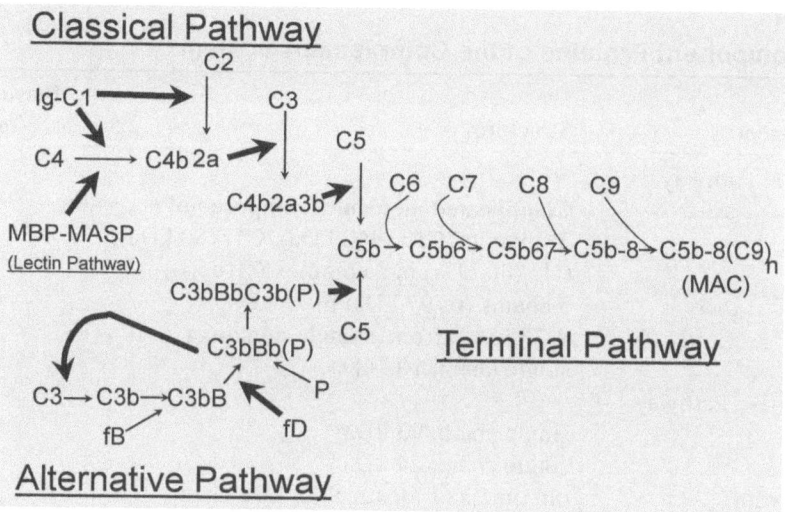

Fig. 1. The complement system and its control. The constituent pathways of the C system and the component proteins are shown. Enzymatic cleavages are represented by thick arrows. The lectin pathway differs from the CP only in that the MBP-MASP complex replaces the C1 complex. Regulators act to inhibit either the enzymes of the activation pathways (activated C1, C3 convertases, C5 convertases) or assembly of the MAC.

2. Activation of C

2.1. The Classical Activation Pathway

The classical activation pathway (CP), so called because it was the first pathway to be described, is triggered by antibody bound to particulate antigen. Many other substances, including components of damaged cells, bacterial lipopolysaccharide, and nucleic acids, can also initiate the CP in an antibody-independent manner.

The CP is initiated by the binding of C1q *(3)*. Activation in vivo involves binding of C1q to aggregated or immune complex bound IgG or IgM antibody. C1 is a large heterooligomeric complex (molecular weight approx 800 kDa) consisting of a single molecule of C1q and two molecules each of C1r and C1s associated noncovalently in a Ca^{2+}-dependent complex. C1q contains no enzymatic activity, but conformational changes occur upon binding of multiple heads of the C1q molecule by aggregates of IgG, which trigger activation of the other components of the C1 complex. IgM is a multivalent molecule and can thus activate C1q efficiently without the need for aggregate formation. The enzymatic activity of the C1 complex is provided by C1r and C1s. These are

Table 1
The Component Proteins of the Complement System

Component	Structure	Plasma conc (mg/L)
Classical Pathway		
C1	Complicated molecule, composed of 3 subunits, C1q (460 kDa), C1r (80 kDa), C1s (80 kDa) in a complex (C1qr$_2$s$_2$)	180
C4	3 chains (α, 97 kDa; β, 75 kDa, γ, 33 kDa); from a single precursor	600
C2	single chain, 102 kDa	20
Alternative Pathway		
fB	single chain, 93 kDa	210
fD	single chain, 24 kDa	2
Properdin	oligomers of identical 53 kDa chains	5
Common:		
C3	2 chains: α, 110 kDa, β, 75 kDa	1300
Terminal Pathway		
C5	2 chains: 115 kDa, 75 kDa	70
C6	single chain, 120 kDa	65
C7	single chain, 110 kDa	55
C8	3 chains: α, 65 kDa, β, 65 kDa γ, 22 kDa	55
C9	single chain, 69kDa	60

The proteins that constitute the classical, alternative, and membrane attack pathways are listed together with their approximate concentration in plasma. Modified from: Morgan, B.P. and Harris, C. L. (1999) *Complement Regulatory Proteins.* Academic, London.

both single-chain molecules of molecular weight 80 kDa, encoded by closely linked genes on chromosome 12 and sharing a high degree of homology. In the presence of Ca^{2+}, C1r and C1s associate with each other to form an elongated C1r$_2$–C1s$_2$ complex, which binds between the globular heads of C1q *(4,5)*. Binding of C1q through the globular heads to Fc regions of aggregated IgG triggers activation through conformational changes that trigger the autoactivation of the proenzyme C1r, a process which involves cleavage at a single site within the molecule.

C1r then activates C1s in the complex, again by cleaving at a single site in the molecule *(4)*. C1s in the activated C1 complex will enzymatically cleave and activate the next component of the CP, C4.

C4 is a large, plasma protein (200 kDa) containing three disulphide-bonded chains (α, β, and γ) *(6,7)*. C4 is encoded by two closely linked genes in the class III region of the major histocompatibility complex (MHC) on the short arm of chromosome 6, which give rise to the two isotypic variants, C4A and C4B *(8,9)*. These variants differ by only six amino acids, but these small changes cause significant differences in function, C4A binding preferentially to amino groups after cleavage and C4B to hydroxyl groups. C1s cleaves plasma C4 at a single site near the amino-terminus of the α-chain, releasing a small fragment, C4a (M_r approx 9 kDa) and, in the process, exposing a reactive thioester group in the α-chain of the large fragment, C4b. The thioester group is crucial to the function of both C4 and C3. In native C4, the thioester is buried deep within a hydrophobic pocket in the molecule, rendering it nonreactive. Once exposed in C4b, the thioester can form covalent amide or ester bonds with exposed amino or hydroxyl groups, respectively, on the activating surface, locking the molecule to the activating surface. The thioester group is extremely labile because of its propensity to inactivation by hydrolysis, restricting C4b binding to the immediate vicinity of the activating C1 complex. Membrane-bound C4b provides a receptor for the next component of the CP, C2.

C2 is a single-chain plasma protein of molecular weight 102 kDa. Membrane-bound C4b expresses a binding site which, in the presence of Mg^{2+} ions, binds C2 and presents it for cleavage by C1s in an adjacent C1 complex. The 70 kDa carboxy-terminal fragment, C2a remains attached to C4b to form the C4b2a complex, the next enzyme in the CP.

C3 is the most abundant (1–2 mg/mL in serum) of the C components, essential for activity of both the CP and AP. It is a large (185 kDa) molecule composed of two chains (α, 110 kDa and β, 75 kDa) held together by disulphide bonds *(10)*. C3 binds noncovalently to C2a in the C4b2a complex and is then cleaved by the C2a enzyme at single site in the α-chain (between *Arg*77 and Ser78), releasing the small fragment, C3a (9 kDa), from the amino-terminus and exposing in the large fragment, C3b, a labile thioester. C3b binds either to the activating C4b2a complex *(11,12)* or to the adjacent membrane. Only C3b bound to the activating C4b2a complex takes any further part in activation, although C3b bound further afield has other important roles in mediating interactions with phagocytic cells. The enzyme so formed, C4b2a3b is the C5 cleaving enzyme (convertase) of the CP.

The next component, C5, is a two-chain plasma protein of 190 kDa molecular weight, structurally related to C3 and C4, but lacking a thioester group. C5 binds noncovalently to a site on C3b in the C4b2a3b convertase and is presented for cleavage by C2a in the complex. Cleavage occurs at a single site (after residue 74) in the α-chain of C5, releasing a small amino-terminal fragment, C5a (approx 10 kDa), and exposing in the larger fragment, C5b, a labile hydrophobic surface-binding site and a site for binding C6.

2.2. The Alternative Activation Pathway

The alternative pathway (AP) provides a rapid, antibody independent route for C activation and amplification on foreign surfaces. C3 is the key component of the AP, but three other proteins, factor B (fB), factor D (fD), and properdin, are also required. FB, a single-chain 93 kDa plasma protein that is closely related to C2, binds C3b in a Mg^{2+}-dependent manner. Binding renders fB susceptible to cleavage by fD, a 26-kDa serine protease present in plasma in its active form, which cleaves fB at a single site, exposing a serine protease domain on the large (60 kDa) fragment, Bb *(13)*. The C3bBb complex thus formed is the C3-cleaving enzyme (C3 convertase) of the AP. Properdin binds and stabilizes the C3bBb complex, extending the lifetime of the active convertase three- or fourfold *(14,15)*. Properdin is a basic glycoprotein made up of oligomers—mainly dimers, trimers, and tetramers—of a 53-kDa monomer *(16)*. The AP requires Mg^{2+} ions for assembly of the C3bBb complex, whereas the CP requires both Mg^{2+} ions (for assembly of the C4b2a complex) and Ca^{2+} ions (for assembly of the C1 complex). This provides a most useful means of distinguishing the two pathways in serum samples. Ethylenediaminetetraacetic acid (EDTA), by chelating both ions, will block both pathways, while ethylene glycol-bis[β-aminoethylether]N,N'-tetraacetic acid (EGTA) (with supplemental Mg^{2+}) chelates only Ca^{2+} and blocks specifically the CP.

Initiation of the AP on a surface occurs spontaneously, the phenomenon of t ickover." C3 in plasma is hydrolyzed to form a metastable $C3(H_2O)$ molecule, which has many of the characteristics of C3b, binds fB in solution, and renders it susceptible to cleavage by fD to form a fluid phase C3 convertase thus formed ($C3(H_2O)Bb$) *(17,18)*. The surface features of many microorganisms and foreign cells favor amplification of the AP (activator surfaces) and rapidly become coated with C3b molecules.

As in the classical activation pathway, bound C3b acts as an essential receptor for C5, permitting the cleavage of C5 by Bb in an adjacent C3bBb complex. The AP C5 convertase comprises two molecules of C3b, one binding Bb in the C3bBb complex and an adjacent (or attached) C3b acting as receptor for C5. The site of cleavage in C5 is identical to that utilized by C2a in the CP convertase.

2.3. The Lectin Pathway

The lectin pathway represents a recently described activation pathway, which provides a second antibody-independent route for activation of C on bacteria and other microorganisms. The key component of this novel pathway is mannan-binding lectin (MBL; also termed mannan-binding protein, MBP), a high-molecular-weight serum lectin made up of multiple copies of a single

32 kDa chain *(2,19)*. MBL binds mannose and N-acetyl glucosamine residues in bacterial cell walls. MBL has a structure similar to that of C1q—a multimeric molecule with globular binding regions and a collagenous stalk, associated with a novel serine protease termed MBL-associated serine protease (MASP), a 100-kDa protein that is homologous with C1r and C1s *(20)*. The MBL-MASP complex activates C4 in a manner analogous to that described for C1, providing a rapid, antibody-independent means of activating the CP on bacteria.

2.4. The Membrane Attack Pathway

The membrane attack pathway involves the noncovalent association of *C5b* with the four terminal C components to form an amphipathic membrane-inserted complex.

While still attached to C3b in the convertase, C5b binds C6, a large (120 kDa) single-chain plasma protein. Binding of C6 stabilizes the membrane binding site in C5b and exposes a binding site for C7, a 110-kDa single-chain plasma protein that is homologous to C6 (*see* below). Attachment of C7 causes release of the complex from the convertase to the fluid phase. The hydrophobic membrane binding site in C5b67 allows the complex to bind tightly with the membrane.

C8 is a complex molecule made up of three chains, α, β, and γ (molecular weights 65, 65, and 22 kDa, respectively). The β-chain in C8 binds C7 in the C5b67 complex and the resulting complex, C5b-8, becomes more deeply buried in the membrane and forms small pores, causing the cell to become slightly leaky *(21)*. C9, a single chain plasma protein (molecular weight 69 kDa), binds the α-chain of C8 in the C5b-8 complex and undergoes a major conformational change from a globular, hydrophilic form to an elongated, amphipathic form which traverses the membrane and exacerbates membrane leakiness. Additional C9 molecules are recruited into the complex to form a pore (the membrane attack complex, MAC), which can cause lysis of the target cell.

3. Regulation of C

The C system is tightly controlled by regulatory proteins present in the plasma and on cell membranes (**Table 2**). The first step of the CP is regulated by C1-inhibitor (C1inh), a serine protease inhibitor that binds activated C1 and removes C1r and C1s from the complex *(22,23)*. C1inh is the only plasma inhibitor of activated C1 and even partial deficiency can result in uncontrolled activation of C in peripheral sites with resultant inflammation (*see* Chapter 12). Control later in the activation pathways is provided by fI, a serine protease which, in the presence of essential cofactors, cleaves C3b and C4b to inactivate the convertases. In plasma, two proteins act as cofactors for fI. fH, a large, single-chain glycoprotein that catalyses fI cleavage of C3b in the AP convertases *(24)*. C4bp is a large, multimeric plasma protein that calalyses cleav-

Table 2
C Regulatory Proteins

Molecule	Structure	Se. conc. (mg/L)	Target
Plasma			
C1inh	single chain, 90 kDa(?)	200	C1
FH	single chain, 90 kDa	450	C3/C5 conv.
FI	2 chains: α, 50 kDa; β, 38 kDa	35	C3/C5 conv.
C4bp	6 or 7 α chains (70 kDa), 1 or 0 β chain (25 kDa)	250	CP C3 conv.
S protein	single chain, 83 kDa	500	C5b-7
Clusterin	2 chains: α, 35 kDa; β, 38 kDa	50	C5b-7
CPN (Anaphylatoxin Inactivator)	dimeric heterodimer, 290 kDa chains, 85 kDa, 50 kDa	30	C3a, C4a, C5a
Membrane			
MCP	single chain, 60 kDa; TM	—	C3/C5 conv.
DAF	single chain, 65 kDa; GPI	—	C3/C5 conv.
CR1	single chain, 200 kDa; TM	—	C3/C5 conv.
CD59	single chain, 20 kDa; GPI	—	C5b-8/C5b-9

RCA, Regulators of Complement Activation gene cluster; TM, transmembrane; GPI, glycosyl phosphatidylinositol; CPN, carboxypeptidase N. Modified from: Morgan, B. P. and Harris, C. L. (1999) *Complement Regulatory Proteins.* Academic, London.

age of C4b in the CP convertase. Both fH and C4bp also inhibit in a second way by acting to break up (decay) the multicomponent convertases, a property termed *decay acceleration (4,25)*. On the membrane, decay accelerating factor (DAF) is a 65-kDa single-chain protein tethered to the membrane by a glycosyl phosphatidylinositol (GPI) anchor. DAF binds to and breaks up the convertase. Membrane cofactor protein (MCP) is a 60-kDa transmembrane protein which, like DAF, binds to the activation pathway convertases but, instead of causing dissociation, acts as cofactor for the cleavage of C4b and C3b by fI, thus irreversibly inactivating the enzyme. **C receptor 1** (CR1) is a large (approx 200 kDa) transmembrane protein that inactivates the activation pathway convertases both by causing dissociation/decay and by acting as a cofactor for cleavage by fI. A unique feature of the AP is the existence of a protein that stabilizes the C3 and C5 cleaving enzymes. Properdin (P) was the first of the components specific to the AP to be discovered *(26)*. It is a large, oligomeric (3, 4, or more identical 56-kDa subunits) plasma protein which binds C3b in the convertase and inhibits the spontaneous and accelerated (fH, DAF, CR1) decay *(27,28)*.

Table 3
Receptors for C components, Fragments, and Complexes

Receptor	Ligand	Characteristics	Distribution
cC1qR	C1q, collagenous region	70 kDa; calmodulin-like	Broad
gC1qR	C1q, globular heads	33 kDa; 80 kDa	Leukocytes; platelets.
CR1 (CD35)	C3b/C4b	180–200 kDa, sc, tm, app. 30 SCRs.	E, B cells, neutrophils, monocytes etc. (broad).
CR2 (CD21)	C3d (and EBV)	145 kDa, sc, tm, 15 or 16 SCRs.	B cells, FDCs, epithelia, glia, ? others?
CR3 (CD11b/18)	iC3b (and matrix)	heterodimer	Myeloid and NK cells
C5aR (CD88)	C5a	40 kDa; 7-tm-spanning	Neutrophils, macrophages, mast cells, muscle, ?others?
C3aR	C3a	app. 60 kDa; 7-tm-sp.	as above
C4aR	C4a	nd	nd
C5b-7R	C5b67	nd	nd

The cellular receptors for C are listed together with their CD asignments (where known) natural ligands, molecular characteristics and cell distribution. Abbreviations: sc, single chain; tm, transmembrane; sp, spanning; SCR, short consensus repeat; E, erythrocyte; FDC, follicular dendritic cell; NK, natural killer cell; nd, not determined. Modified from: Morgan, B. P. and Harris, C. L. (1999) *Complement Regulatory Proteins.* Academic, London.

The membrane attack pathway is tightly regulated by inhibitors present in the fluid phase and on membranes. The hydrophobic membrane binding site in the fluid-phase C5b-7 complex is the target of S-protein (vitronectin) and clusterin, abundant serum proteins which, among their many roles, help regulate C activation. On the membrane, CD59 antigen (CD59), a 20-kDa GPI-anchored molecule, binds to C8 in the C5b-8 complex and blocks incorporation of C9, thereby preventing formation of the lytic MAC (*29,30*).

4. Receptors for C Components and Fragments

Cells express surface receptors specific for *C1q*, for the large fragments of C3 and C4 generated during C activation and for the small fragments of C3, C4, and C5 (**Table 3**). Receptors for C1q (C1qR) were first described in the early 1980s (*31*). Several different receptors for C1q, binding either the collagenous stalk or the globular head, have now been described, although their precise roles remain to be ascertained (*32–34*). Multiple receptors also exist for the large products of cleavage of C4 (C4b) and C3 (C3b) generated during C activation. **C receptor 1** (CR1; CD35) has dual roles as C regulator and C receptor. The ligands for CR1 are C3b and C4b, which bind to separate sites on the large CR1 molecule *(35)*. CR1 on erythrocytes plays a key role in the transport of immune complexes. **C receptor 2** (CR2; CD21) is the receptor for C3d, the surface-bound fragment, which is the end-product of the cleavage of C3b by fI and serum proteases *(36)*. CR2 does not function as a C regulator. **C receptor 3** (CR3; CD11b/CD18) is the receptor for iC3b and also binds numerous extracellular matrix proteins. It is a member of the β-2 integrin family of cell adhesion molecules and, like other members of this family, is a heterodimer composed of one molecule of the common integrin β-chain, CD18, and one molecule of a specific α-chain, which in CR3 is CD11b *(37,38)*. **C receptor 4** (CR4; CD11c/CD18), like CR3, is a β-2 integrin with wider roles in cell adhesion.

The small anaphylactic peptides C5a and C3a generated during C activation express important biological activities. Each peptide is 74–77 aminoacids in length and is highly cationic. The presence of specific and distinct membrane receptors for C5a and C3a on neutrophils and macrophages was first demonstrated by classical methods using labeled ligands *(39,40)*. The C5a receptor (C5aR; CD88) was cloned and shown to be a member of the 7-transmembrane-spanning receptor family *(41)*. The precise distribution of the receptor is still unclear. The receptor for C3a (C3aR) was finally cloned in 1996 *(42)*. C3aR is also a member of the 7-transmembrane-spanning receptor family and is highly homologous with C5aR, the major difference being a large extracellular loop, absent in C5aR.

5. Deficiencies of C

Deficiencies of almost every C protein and regulator have been described and more detailed accounts of the various C deficiencies can be found in several recent reviews *(43–45)*. A description of the considerations and methods applied to the identification of C deficiencies is provided by A. Orren in Chapter 11 and deficiency of C1 inhibitor is described by E. Hack in Chapter 12.

Deficiencies of components of the CP (C1, C4, or C2) are associated particularly with an increased susceptibility to immune complex disease, a consequence of the failure of immune complex solubilization. The frequency and severity of disease is greatest with deficiencies of one of the subunits of C1 (C1q, C1r, C1s), closely followed by total C4 deficiency, each giving rise to a severe immune complex disease, which closely resembles systemic lupus erythematosus (SLE). Subtotal deficiency of C4 is common, because of the extremely high frequency of null alleles at both the C4A and C4B loci, but total C4 deficiency is very rare. Deficiency of C2 is the most common homozygous C deficiency in Caucasoids, but causes much less severe disease; it appears that deposition of the early components, C1 and C4, provides some solubilization of immune complexes.

C3 is an essential component of both activation pathways and is vital for efficient opsonisation of bacteria. Deficiency thus causes a marked susceptibility to bacterial infections. Immune complex disease is not a common finding in C3 deficiency, although solubilization of preformed immune complexes is severely compromised in the absence of C3. Individuals with C3 deficiency run a stormy course through childhood, but with prompt therapy of infections can survive. In adulthood, the number and severity of infections is much reduced as other arms of the immune system take up the challenge. Deficiencies of the regulators fI and fH cause a secondary deficiency of C3 and present with similar symptoms.

Deficiencies of components of the AP are rare and do not predispose to immune complex disease or pyogenic infections. A few individuals deficient in fD have been described, all of whom have presented with recurrent *Neisseria* infections, usually meningococcal meningitis. Very recently, two cases of fH deficiency have been described, again presenting with meningococcal disease. Deficiency of the positive regulator, properdin, is the commonest disorder of the AP.

Deficiencies of terminal pathway components (C5, C6, C7, C8, or C9) also cause susceptibility to infection with organisms of the genus *Neisseria*. Deficiency of C6 is the second most common C deficiency among Caucasians and is frequently associated with meningococcal meningitis and systemic infection with the meningococcus. C9 deficiency, rare in Caucasians, is by far the most frequent C deficiency in Japan with an incidence approaching 1 in 1000 *(46)*.

6. C in Inflammatory Disease

The C system contributes to tissue damage in a large number of autoimmune diseases. Autoantibodies and immune complexes deposit in the affected organs where they trigger activation of C and exacerbate inflammation. In many autoimmune diseases, from the organ specific (e.g., autoimmune thyroid disease) to the disseminated (e.g., SLE), C deposition can be detected in the affected tissues and products of C activation are found in the plasma (methods for detection detailed in Chapters 8 and 7, respectively). In all these autoimmune diseases, C is just one of several factors that contribute to pathogenesis. In SLE, autoantibodies are present that recognize DNA and other components of normal cells. Following any stimulus to cell death, these components will be released and immune complexes will form. Immune complexes deposit in capillary beds, particularly in skin and kidney, where they activate C to cause inflammation and further tissue destruction (*see* Chapter 15). A vicious cycle is triggered in which more cell killing drives the production of more immune complexes, which in turn exacerbates C activation and tissue destruction.

References

1. Turner, M. W. (1991) Deficiency of mannan binding protein—a new complement deficiency syndrome. *Clinic. Experiment. Immunol.* **86,** 53–56.
2. Reid, K. B. and Turner, M. W. (1994) Mammalian lectins in activation and clearance mechanisms involving the complement system. *Springer Sem. Immunopathol.* **15,** 307–326.
3. Loos, M. (1988) "Classical" pathway of activation, in *The Complement System* (Rother, K. and Till, G. O., eds.), Springer, Berlin, pp. 136–154.
4. Reid, K. B. (1986) Activation and control of the complement system. *Essays in Biochem.* **22,** 27–68.
5. Reid, K. B. and Day, A. J. (1989) Structure-function relationships of the complement components. *Immunol. Today* **10,** 177–180.
6. Schreiber, R. D. and Muller-Eberhard, H. J. (1974) Fourth component of human complement: description of a three polypeptide chain structure. *J. Exp. Med.* **140,** 1324–1335.
7. Janatova, J. and Tack, B. F. (1981) Fourth component of human complement: studies of an amine-sensitive site comprised of a thiol component. *Biochemistry* **20,** 2394–2402.
8. Campbell, R. D., Dunham, I., and Sargent, C. A. (1988) Molecular mapping of the HLA-linked complement genes and the RCA linkage group. *Experiment. Clinic. Immunogenet.* **5,** 81–98.
9. Campbell, R. D. (1988) The molecular genetics of components of the complement system. *Baillieres Clinic. Rheumatol.* **2,** 547–575.
10. Lambris, J. D. (1988) The multifunctional role of C3, the third component of complement. *Immunol. Today* **9,** 387–393.
11. Kozono, H., Kinoshita, T., Kim, Y. U., Takata-Kozono, Y., Tsunasawa, S., Sakiyama, F., et al. (1990) Localization of the covalent C3b-binding site on C4b within the complement classical pathway C5 convertase, C4b2a3b. *J. Biolog. Chem.* **265,** 14,444–14,449.
12. Ebanks, R. O., Jaikaran, A. S., Carroll, M. C., Anderson, M. J., Campbell, R. D., and Isenman, D. E. (1992) A single arginine to tryptophan interchange at beta-chain residue 458 of human complement component C4 accounts for the defect in classical pathway C5 convertase activity of allotype C4A6. Implications for the location of a C5 binding site in C4. *J. Immunol.* **148,** 2803–2811.
13. Gotze, O. (1986) The alternative pathway of activation, in *The Complement System* (Rother, K. and Till, G. O., eds.), Springer, Berlin, pp. 154–168.
14. Weiler, J. M., Daha, M. R., Austen, K. F., and Fearon, D. T. (1976) Control of the amplification convertase of complement by the plasma protein beta1H. *Proc. Natl. Acad. Sci. USA* **73,** 3268–3272.
15. Fearon, D. T., Daha, M. R., Weiler, J. M., and Austen, K. F. (1976) The natural modulation of the amplification phase of complement activation. *Transplant. Rev.* **32,** 12–25.
16. Minta, J. O. and Lepow, I. H. (1974) Studies on the subunit structure of human properdin. *Immunochemistry* **11,** 361–368.

17. Lachmann, P. J. and Hughes-Jones, N. C. (1984) Initiation of complement activation. *Springer Sem. Immunopathol.* **7,** 143–162.
18. Law, S. K. and Dodds, A. W. (1990) C3, C4 and C5: the thioester site. *Biochem. Soc. Trans.* **18,** 1155–1159.
19. Holmskov, U., Malhotra, R., Sim, R. B., and Jensenius, J. C. (1994) Collectins: collagenous C-type lectins of the innate immune defense system. *Immunol. Today* 67–74.
20. Matsuhita, M. and Fujita, T. (1992) Activation of the classical complement pathway by mannose-binding protein in association with a novel *C1s*-like serine protease. *J. Exp. Med.* **176,** 1497–1502.
21. Tamura, N., Shimada, A., and Chang, S. (1972) Further evidence for immune cytolysis by antibody and the first eight components of complement. *Immunology* **22,** 131–140.
22. Davis, A. E. (1988) C1 inhibitor and hereditary angioneurotic edema. *Annu. Rev. Immunol.* **5,** 595–628.
23. Davis, A. E. (1989) Hereditary and acquired deficiencies of C1 inhibitor. *Immunodef. Rev.* **1,** 207–226.
24. Vik, D. P., Munoz-Canoves, P., Chaplin, D. D., and Tack, B. F. (1990) Factor H. *Curr. Topics Microbiol. Immunol.* **153,** 147–162.
25. Gigli, I., Fujita, T., and Nussenzweig, V. (1979) Modulation of the classical pathway C3 convertase by the plasma proteins C4 binding protein and *C3b* inactivator. *Proc. Natl. Acad. Sci. USA* **76,** 6596–6600.
26. Pillemer, L., Blum, L., Lepow, I. H., Ross, O. A., Todd, E. W., and Wardlaw, A. C. (1954) The properdin system and immunity. I. demonstration of a new serum protein, properdin, and its role in immune phenomena. *Science* **120,** 279–285.
27. Smith, C. A., Pangburn, M. K., Vogel, C-W., and Muller-Eberhard, H. J. (1984) Molecular architecture of human properdin, a positive regulator of the alternative pathway of human complement. *J. Biol. Chem.* **259,** 4582–4588.
28. Pangburn, M. K. (1986) The alternative pathway, in *Immunobiology of the complement system* (Ross, G. D., ed.), Academic, New York, pp. 45–62.
29. Lachmann, P. J. (1991) The control of homologous lysis. *Immunol. Today* **12,** 312–315.
30. Davies, A. and Lachmann, P. J. (1993) Membrane defence against complement lysis the structure and biological properties of CD59. *Immunol. Res.* **12,** 258–275.
31. Andrews, B. S., Shadforth, M., Cunningham, P., and Davis, J. S. (1981) Demonstration of a *C1q* receptor on the surface of human endothelial cells. *J. Immunol.* **127,** 1075–1080.
32. Ghebrehiwet, B. (1989) Functions associated with the C1q receptor. *Behring Inst. Mitt.* **84,** 204–215.
33. Sim, R. B. and Malhotra, R. (1994) Interactions of carbohydrates and lectins with complement. *Biochem. Soc. Trans.* **22,** 106–111.
34. Eggleton, P., Gehebrehewit, B., Sastry, K. N., Coburn, J. P., Zaner, K. S., Reid, K. B., and Tauber, A. I. (1995) Identification of a gC1q-binding protein (gC1q-R) on the surface of human neutrophils. Subcellular localisation and binding properties in comparison with the cC1q-R. *J. Clin. Invest.* **95,** 1569–1578.

35. Fearon, D. T. and Wong, W. W. (1983) Complement ligand-receptor interactions that mediate biological responses. *Ann. Rev. Immunol.* **1,** 243–271.
36. Fearon, D. T., Klickstein, L. B., Wong, W. W., Wilson, J. G., Moore, F. D., Jr., Weis, J. J., Weis, et al. (1989) Immunoregulatory functions of complement: structural and functional studies of complement receptor type 1 (CR1; CD35) and type 2 (CR2; CD21). *Progr. Clin. Biolog. Res.* **297,** 211–220.
37. Rothlein, R. and Springer, T. A. (1985) Complement receptor type three-dependent degradation of opsonized erythrocytes by mouse macrophages. *J. Immunol.* **135,** 2668–2672.
38. Larson, R. S. and Springer, T. A. (1990) Structure and function of leukocyte integrins. *Immunolog. Rev.* **114,** 181–217.
39. Chenoweth, D. E. and Goodman, M. G. (1983) The C5a receptor of neutrophils and macrophages. *Agents & Actions—Suppl.* **12,** 252–273.
40. van Epps, D. E. and Chenoweth, D. E. (1984) Analysis of the binding of fluorescent C5a and C3a to human peripheral blood leukocytes. *J. Immunol.* **132,** 2862–2867.
41. Gerard, N. P. and Gerard, C. (1991) The chemotactic receptor for human C5a anaphylatoxin. *Nature* **349,** 614–617.
42. Ames, R. S., Li, Y., Sarau, H. M., Nuthulaganti, P., Foley, J. J., Ellis, C., et al. (1996) Molecular cloning and characterization of the human anaphylatoxin C3a receptor. *J. Biol. Chem.* **271,** 20,231–20,234.
43. Morgan, B. P. and Walport, M. J. (1991) Complement deficiency and disease. *Immunol. Today* **12,** 301–306.
44. Colten, H. R. and Rosen, F. S. (1992) Complement deficiencies. *Ann. Rev. Immunol.* **10,** 809–834.
45. Figueroa, J., Andreoni, J., and Densen, P. (1993) Complement deficiency states and meningococcal disease. *Immunolog. Res.* **12,** 295–311.
46. Fukumori, Y., Yoshimura, K., Ohnoki, S., Yamaguchi, H., Akagaki, Y., and Inai, S. (1989) A high incidence of C9 deficiency among healthy blood donors in Osaka, Japan. *Int. Immunol.* **1,** 85–89.

2

Purification of Complement Components, Regulators, and Receptors by Classical Methods

Carmen W. van den Berg

1. Introduction

1.1. Overview

The aim of this chapter is to describe methods for purification of the individual complement (C) components using classical chromatography methods available in most biochemistry laboratories. None of these methods require the large amounts of specific antibodies needed for the popular and rapid immunoaffinity methods to be described in Chapter 3. A further advantage of classical chromatography is that the methods described in this chapter for human C can easily be adapted for purification of C components of other species. Indeed, some of the methods described here were originally developed for the purification of animal components, but proved also to be suitable for human components. One precautionary note when adapting methods to C components of another species is that there is occasionally species incompatibility in that some components will not function in combination with components of another species. Functional assays may, therefore, require modification. It should also be noted that antisera and monoclonal antibodies against a C component frequently do not crossreact with that component from other species.

An important consideration when purifying C components is the freshness of the serum or plasma used as raw material. It is important to start whenever possible with fresh plasma because this will increase the yield of active components. In stored plasma, C components will have undergone proteolysis or other forms of inactivation to varying degrees. C activation and consumption of C components is also a potential problem, particularly if serum is used as source. To prevent C activation, the use of plasma is preferred, and plasma

From: *Methods in Molecular Biology, vol. 150: Complement Methods and Protocols*
Edited by: B. P. Morgan © Humana Press Inc., Totowa, NJ

should be stored at –70°C in the presence of protease inhibitors. To minimize C activation or degradation of proteins by enzymes in the serum/plasma during the purification procedure, all the purification steps are carried out at 4°C unless mentioned otherwise. Even when rapid methods such as fast protein liquid chromatography (FPLC) are used to minimize the run time, activation can occur if runs are performed at room temperature and it is advisable to run the FPLC in a 4°C cold room. Protease inhibitors should also be included during purification, particularly in the early stages when plasma enzymes are present in abundance. Of the protease inhibitors, NPGB (p-nitro phenyl p'-guanidino-benzoate) is a particularly good inhibitor of C1r and C1s activation; it is dissolved as a 100× concentrated stock solution (100 mM) in dimethyl formamide (DMF) and used fresh. Phenyl methyl sulfonyl fluoride (PMSF), a broad-range protease inhibitor, is dissolved as a 100× stock solution (100 mM) in isopropanol. Many other protease inhibitors are soluble in water and can be directly dissolved in the required buffer. Diisopropyl fluoro phosphate (DFP) is suggested in some published purification methods, however this agent is highly toxic and can often replaced by PMSF. All the above protease inhibitors are toxic to varying degrees and should be handled with extreme caution. Ethylene diamine tetraacetic acid (EDTA) can also be added to chelate calcium and magnesium in order to prevent C activation and to inhibit activation or degradation by metalloproteases in the serum/plasma, however, some multichain complexes, like C1, require divalent cations and will disassociate in the presence of EDTA.

Plasma and serum to be used for preparation of C components and column fractions at various stages of the purification should be stored at –70°C. However, repeated freeze-thawing will also lead to C activation. C3, C4, and C5 are particularly susceptible to damage or activation by freeze-thaw and fractions containing these components are better stored at 4°C for short periods of time.

1.2. Purification Steps—General Considerations

Most purification methods for C components involve the use of an initial fractionation step to precipitate the protein of interest, followed by standard chromatography using multiple columns. Some methods have been developed that rely on the affinity of the component for its substrate/ligand. The first step in purification usually involves a crude enrichment of the protein required by fractional precipitation. For some components this involves a euglobulin precipitation (i.e., low salt) or salting out using ammonium sulfate or sodium sulfate. More common in recent methods are procedures involving fractionated precipitation using polyethylene glycol (PEG). PEG separates molecules roughly according to molecular weight; the larger molecules precipitating at lower PEG concentrations. PEG is supplied with different average chain lengths (shown as a suffix indicating average molecular weight) and a longer PEG species (e.g., PEG 6000) will precipitate a given protein at a lower percentage than a shorter PEG (e.g., PEG 3350). It is, thus, extremely important, when following a published protocol, to use the specified PEG species for precipitation. PEG precipitation has advantages over euglobulin or salt precipitation in that it is much faster and generally less denaturing for the protein, but the main advantage is that the dissolved precipitate can be resuspended in a buffer required for the first column step and can be applied directly onto ion-exchange columns without the requirement for prior dialysis to remove salt. Fractionation by PEG precipitation is more efficient and selective when the plasma or serum is first ten times diluted in buffer. However, this is not practical for large volumes of serum. To prevent blockage of columns, redissolved PEG precipitates must always be filtered (0.2 μm) or spun at high speed (10,000g, 15 min) to remove any insoluble material.

Columns and column matrices for the various chromatography steps can be obtained from a variety of suppliers including Pharmacia, Bio-Rad, Sigma, and Calbiochem. It is often possible to replace matrices described in the original methods with other media providing the same chemistry for separation and substitute low-resolution columns with high-performance columns. The use of high-resolution, fast-flow columns operating at higher pressures makes purification times shorter and thus reduces the chance of C activation. FPLC columns are particular useful for purification of C components from small amounts of serum and methods described here using large amounts of serum can easily be scaled down to use FPLC columns with increased efficiency and speed of purification, often requiring fewer chromatography steps.

Gel filtration is often used as a final "polishing" step to remove residual contaminants or aggregates and is rarely of use early in a purification protocol. For application onto gel filtration columns, proteins usually have to be concentrated. This can be carried out on an Amicon ultrafiltration system or any one of several other commercially available ultrafiltration systems. Ultrafiltration also permits rapid buffer exchange and can eliminate the need for dialysis. Care should be take to choose a filter with the right molecular weight cutoff for the protein of interest; occasionally, proteins with a reported molecular weight higher than the stated cutoff will still go through the filter; this must be assessed on an individual basis. For most C components, a filter with a cutoff of 30 kDa offers a good compromise. Care should also be taken with some proteins that are sticky and will bind to some filters, resulting in considerable losses during the concentration steps.

Most of the methods described in this chapter are designed to obtain biochemically pure proteins (>90% pure). When all that is required for a particular purpose is a functionally pure protein (a preparation that still contain some contaminants, but only contaminants that do not interfere with the aim of the experiment for which the protein is being purified) some of the columns can be omitted. When biochemically pure components are required and the yield is not very important, it is best to pool fractions tightly, discarding fractions on the edges of the protein peak which may contain difficult to remove contaminants. Some common contaminants can be removed by inclusion of a specific clean-up step; albumin can be removed by passage over Cibacron Blue Sepharose and immunoglobulin by passage over protein-A or protein-G columns.

Most methods described in this chapter are for the purification of a single component. Methods have been described for the purification of multiple complement component from a single batch of plasma *(1,2)*. Although these multicomponent methods are economical in usage of raw materials, for most workers they are unnecessary and far too complex. Nevertheless, they do highlight the fact that several components can be obtained from a single pool of plasma without too much difficulty. Fractions obtained during the purification of a specific component that do not contain the component of interest may be frozen at −70°C and used at a later stage to purify other components.

Screening fractions for the component of interest after each of the purification steps is perhaps the most challenging aspect of a purification protocol. This can best be carried out using a functional hemolytic assay. The use of functional assays also ensures that the functionally active component and not an inactive product is being purified. Throughout this chapter, the aim is to suggest the simplest screening methods possible using sera deficient in or specifically depleted of a single component. Sera obtained from patients (or animals) deficient in individual components are now quite widely available; methods for the generation of sera depleted of specific components are described in Chapter 4. Some C deficient and depleted sera can be obtained from commercial sources (*see* Appendix).

2. Materials

2.1. Buffers

1. Phosphate-buffered saline (PBS): supplied in tablet form by Oxoid Ltd.
2. Veronal-buffered saline (VBS^{2+}): supplied in tablet form by Oxoid Ltd.
3. Tris-buffered saline (TBS): 50 mM Tris-HCl, 150 mM NaCl, pH 7.5.
4. Various specific buffers for different column chemistries, including phosphate-, Tris- and barbitone-based buffers, detailed with the specific method.

2.2. Column Matrices

1. Matrices for gel filtration: Sepharose, Sephadex, and sephacryl matrices, precise type chosen depending on M$_r$ of protein of interest (all from Amersham Pharmacia).

2. Matrices for ion exchange: diethylaminoethyl (DEAE) Sepharose and carboxymethyl (CM) sepharose (Amersham Pharmacia).
3. Prepoured columns for FPLC gel filtration and ion exchange (all from Amersham Pharmacia).

2.3. Starting Materials

1. For plasma proteins, a source of plasma in sufficient quantities for the planned purification. Fresh EDTA plasma is best for most C components, although fresh serum works better for others. Expired plasma (available from Blood Banks) is suitable for purification of some C components, although some degree of proteolysis is often seen.
2. For erythrocyte-expressed C regulators or receptors, a source of erythrocytes in sufficient quantities for the planned purification. Expired erythrocytes are usually adequate. Platelets are often available from blood banks in large quantities and offer an alternative source for membrane C regulators and receptors.

2.4. Equipment

1. Pumps, fraction collectors, and absorbance monitors for standard protein purification system (available from many manufactureres, including Bio-Rad and Amersham, Pharmacia).
2. Automated or semiautomated chromatography systems are also widely available and can simplify preparations.
3. Spectrophotometer capable of reading absorbance in ultraviolet (UV) and visible ranges.
4. A cold room or cold cabinet in which to perform chromatographic steps.
5. Centrifuges: Ranging from benchtop microfuges to large, floor-standing refrigerated centrifuges capable of processing 5–10 L of fluid at once.
6. Ultrafiltration system for concentrating/dialyzing samples between chromatography steps. We use an Amicon ultrafiltration cell (Amicon, Inc.); numerous other systems are available.

3. Methods

3.1. Purification of Classical Pathway Components

3.1.1. Purification of C1

C1; M_r 740 kDa. Heteromeric protein, $C1qC1r_2C1s_2$.
C1q: 450 kDa: 6 chains each of: A: 29 kDa, B: 27 kDa, C: 22 kDa. Serum conc.: 180 µg/mL.
C1r: 85 kDa; activated C1r: 56 kDa and 35 kDa; serum conc.: 100 µg/mL.
C1s: 85 kDa activated C1s: 56 kDa, 27 kDa; serum conc.: 80 µg/mL.

3.1.1.1. FUNCTIONALLY PURE C1:

Functionally pure C1 can be obtained by euglobulin precipitation of serum. The usual starting material for this procedure is fresh citrated plasma (plasma anticoagulated with EDTA or heparin will not do).

1. Fresh citrated plasma is recalcified by addition of $CaCl_2$ from 1 M stock to 20 mM and allowed to clot for 2 h at room temperature.
2. The clot is removed by centrifugation and the protease inhibitor NPGB added to a final concentration of 1 mM.
3. The serum is dialyzed overnight at 4°C against a 10-fold excess of 10 mM NaBarbitone buffer pH 7.4 containing $CaCl_2$ (5 mM) and NPGB (0.1 mM). In order to isolate C1q, dialyze in this step against 5 mM EDTA alone and proceed as detailed below.
4. The fine euglobulin precipitate which forms in the serum is harvested by centrifugation at 10,000g for 30 min at 4°C and dissolved in VBS containing 5 mM Ca^{2+}, 0.1 mM NPGB, and 0.1 mM PMSF.
5. As an optional extra step, the redissolved euglobulin preparation can be subjected to gel filtration on a Sepharose 6B (or similar) column equilibrated in the above buffer; the C1 elutes early.
6. The "functionally pure" C1 can be is stored in small aliquots at –70°C and should not be subjected to freeze-thaw cycles.

3.1.1.2. "Biochemically Pure" C1

Isolation of "biochemically pure" C1 and its subcomponents is most efficiently achieved using methods based on the affinity of C1/C1q for aggregated IgG *(3)*.

The isolation of unactivated "precursor" C1 free of "activated" C1 in which C1r and C1s are cleaved is difficult and reliant on the speed of purification and attention to maintaining low temperatures during purification.

1. A human IgG-Sepharose column (55 mL: 3.5 mg IgG/mL Sepharose) is made by incubating human IgG (Sigma) with CNBr-activated sepharose (Pharmacia) according to the manufacturer's instructions.
2. The column matrix is saturated with rabbit IgG by repeated passage of rabbit antihuman IgG antiserum containing 10 mM EDTA (approx 50 mL) over the column.
3. The column is washed with PBS containing 1 M NaCl to remove unbound protein and equilibrated with VBS.
4. Human serum (100 mL), obtained from citrated plasma as described above, is diluted 1:1 with VBS containing 5 mM $CaCl_2$ and 1 mM NPGB and applied to the column.
5. The column is washed with three column volumes of the same buffer and C1 eluted in the same buffer containing 1 M NaCl. If only C1q is required, load the IgG-column with serum in the presence of 10 mM EDTA and elute with 1 M NaCl. For purification of separate components of C1, the protocol up to **step 4** is identical to that described above, then as below.
6. The column is washed with three column volumes of VBS containing 5 mM $CaCl_2$ and 1 mM NPGB.
7. C1r and C1s are then eluted in VBS containing 10 mM EDTA and 1 mM NPGB.
8. C1q is then eluted in the same buffer containing 1 M NaCl.

9. C1q can be further purified by dialysis into 20 m*M* HEPES, 60 m*M* NaCl, 10 m*M* EDTA, pH 7.8, application to a Mono S FPLC column equilibrated in the same buffer and elution with a 20-mL linear NaCl gradient from 0.06 to 1 *M* NaCl *(4)*.

10. To separate C1r from C1s, the eluted material from **step 5** is dialyzed against 20 m*M* NaPhosphate, 2 m*M* EDTA, pH 7.4, applied to a MonoQ column equilibrated in the same buffer and eluted using a 20-mL linear NaCl gradient from 0–1 *M (5)*. C1r elutes before C1s.
   See **Note 1**.

3.1.1.3. FUNCTIONAL ASSAYS FOR C1

3.1.1.3.1. Hemolysis Method

1. Human serum is depleted of *C1* (NHS-C1) by precipitation with 3.5% PEG 4000 (30 min on ice) and the C1-containing precipitate removed by centrifugation (15 min, 10,000*g*, 4°C) *(6)*.
2. ShEA (50 µL 2%) are incubated in wells of a microtitre plate with 50 µL NHS-C1 diluted 1/10 in VBS and 50 µL of the test fractions for 30 min at 37°C.
3. Incubate for 30 min at 37°C, spin, and read the absorbance of the supernatant at 415 nm.
4. Fractions restoring C1 hemolytic activity are pooled for further use.

3.1.1.3.2. Esterolytic Assay: Esterolytic activity: functional activities of C1r and C1s can be measure using the chromogenic substrate ZLNE [N α carbobenzoxy-L-lysine-p-nitrophenyl ester (Sigma)] or S-2314.

1. Place 60-µL aliquots of fractions containing C1r or C1s in the wells of a 96-well plate and add 1 µL stock ZLNE (3 mg/mL in 90%(v/v) acetonitrile in water, made fresh immediately before use).
2. Incubate 30 min 37°C and measure release of p-nitrophenol spectrophotometrically at 340 nm.

3.1.2. Purification of MBL, MASP-1, and MASP-2

Mannose-binding lectin (MBL): M_r 750 kDa; serum conc.: 2 ng–10 µg/mL.
MBL associated serine protease (MASP)-1: M_r 90 kDa; upon activation: 65 and 25 kDa. Serum conc.: 6 µg/mL.
MASP-2: 76 kDa; upon activation: 52 and 31 kDa; serum conc. not known, but much lower than MASP-1.

Purification of MBL is based on carbohydrate affinity chromatography and the sequential use of different sugars to remove contaminating anticarbohydrate antibodies from the MBL preparation *(7)*.

1. Serum (1 L) is brought to 7% w/v PEG3500, stirred for 2 h at 4°C, and spun for 15 min at 12,000*g*.
2. The PEG precipitate is dissolved in 400-mL TBS containing 0.05% Tween20, 20 m*M* $CaCl_2$, pH 7.8 (TBS-TCa) and mixed for 2 h at 4°C with 50 mL

mannose Sepharose (mannose coupled to Sepharose 4B), equilibrated in the same buffer.

3. The matrix is placed in a column, washed with 500 mL of the same buffer, and eluted with buffer containing no added Ca^{2+} and 10 mM EDTA.

4. The eluate is made 40 mM with $CaCl_2$, applied to a 3-mL maltose-Sepharose (maltose coupled to Sepharose 4B) column, washed with TBS-TCa buffer and eluted with 100 mM GlcNAc (N-acetyl-D-glucosamine) in TBS-TCa.

5. If further purification is required, the MBL-positive pool is dialyzed into 50 mM Tris/100-mM NaCl, 2-mM EDTA, pH 7.8, applied to a MonoQ FPLC column equilibrated in the same buffer and eluted with a 40-mL linear NaCl gradient from 0.1 to 0.6 M.

6. MBL-positive fractions are pooled and applied to a Superose 6 FPLC column; positive peak contains MBL and bound MASP-1/2.

See **Note 1**.

3.1.3. Purification of C4

C4: M_r 189 kDa: α: 89 kDa, β: 72 kDa, γ: 32 kDa;
Activated: C4a: 9 kDa, C4α:80 kDa: serum conc. 200–600 μg/mL.

1. Supplement 1 L of citrated plasma with 5 mM benzamidine (to precipitate vitamin K-dependent proteases) and 0.5 mM PMSF; add solid PEG6000 to a final conc. of 5%, stir for 30 min on ice, spin 15 min 6000g 4°C. Add solid PEG6000 to a final conc. of 8%, stir 30 min on ice, and spin 15 min 6000g 4°C (*8*).

2. Dissolve the pellet in 10 mM Tris, 10 mM EDTA, 150 mM NaCl, pH 8.0, and apply to a Q-Sepharose column (5 cm × 20 cm) equilibrated in the same buffer. Elute with a 1 L linear NaCl gradient from 0.15 to 0.5 M.

3. Pool positive fractions, dialyze against 10 mM Tris, 10 mM EDTA, 150 mM NaCl, pH 8.0, and apply to a MonoQ FPLC column equilibrated in the same buffer. Elute with a 30-mL linear NaCl gradient from 0.15 to 0.5 M.

4. Pool positive fractions, concentrate in Amicon ultrafiltration cell and gel filter on a Superose 12 FPLC column equilibrated in PBS. Pool and concentrate positive fractions.

3.1.3.1. FUNCTIONAL ASSAY FOR C4

1. Generate C4-depleted guinea pig serum (R4) by incubating fresh guinea pig serum (8.5 mL) with 5% NH_4OH (1.5 mL) at 37°C for 45 min. Immediately adjust pH to 7.4 and store in aliquots at –70°C. If available, C4-deficient guinea pig serum can be used in place of R4.

2. To 50 μL 2% ShEA in VBS, add 50 μL C4-depleted serum (1/5 in VBS) and 50 μL of test fraction for C4.

3. Incubate for 30 min at 37°C, spin, and read the absorbance of the supernatant at 415 nm.

4. Fractions restoring C4 hemolytic activity are pooled for further use.

3.1.3.2. GENERATION OF C4 FRAGMENTS

C4a and C4b can be generated by incubation of purified C4 with activated C1s.

1. To purified C4 (1 mg/mL in PBS) add activated C1 (2% w/w) and incubate for 1–2 h at 37°C.
2. Separate C4a and C4b by gel filtration on a Sephadex G-100 Superfine (or similar) column equilibrated in 10 mM NaPhosphate, 1 M NaCl, and pH 7.2.
3. Generate C4c and C4d by incubation of C4b (1 mg/mL in PBS) with factor I (2% w/w) and C4bp (10% w/w) at 37°C for 1 h.
4. Separate C4c and C4d from uncleaved C4b by gel filtration on a Sephadex G-150 (or similar) column in 10 mM NaPhosphate, 1 M NaCl, pH 7.2.
5. Generate iC4b by incubation of C4b with 500 mM methylamine, pH 8.0 for 1 h at 37°C. Generate C4$_{MA}$ by incubation of C4 with methylamine as above. C4b can be separated from iC4b and C4$_{MA}$ from C4 by loading on a MonoQ column in 20-mM Tris, 2 mM EDTA, 150 mM NaCl, pH 7.4 and eluting with a 20-mL linear NaCl gradient from 0.15 to 0.5 M.

3.1.4. Purification of C2 (9)

C2: M$_r$: 110 kDa, activated C2a: 74 kDa, C2b: 34 kDa, serum conc. 25 µg/mL.

1. Dialyze 1 L of serum overnight against 5 mM CaCl$_2$, 1 mM Benzamidine, and spin 30 min 23,000g.
2. Add 10 mL 200 mM EDTA and 50 mL 400 mM NaPhosphate, 40 mM EDTA, pH 6.0 to the supernatant, and adjust pH to 6.0. Apply to CM-Sephadex C-50 (8 cm × 10 cm) column equilibrated in 100 mM NaPhosphate, 1 mM benzamidine, pH 6.0, and elute with a 2-L linear phosphate gradient from 0.1 to 0.25 M NaPhosphate, and pH 6.0.
3. To the C2-positive fractions, add (NH$_4$)$_2$SO$_4$ to 50% saturation (291 g/L), stir 1 h 4°C, and centrifuge 23,000g for 30 min at RT.
4. To the supernatant, add (NH$_4$)$_2$SO$_4$ to 75% saturation (159 g/L), stir 1 h at RT and spin 23,000g, 90 min at RT.
5. Resuspend pellet in 50 mL, 5 mM NaBarbitone, 0.5 mM CaCl$_2$, 2 mM MgCl$_2$, 40 mM NaCl, pH 8.5, and gel filter on Sepharose 4B (or similar) column (4 cm × 30 cm) equilibrated in the same buffer.
6. Pool and concentrate C2-positive fractions.

3.1.4.1. FUNCTIONAL ASSAY FOR C2

C2 and fB are the most thermosensitive of the C components and can be specifically depleted from serum by careful heating.

1. Serum (1 mL) is placed in a glass tube preheated in a 56°C water bath and incubated at 56°C for precisely 6 min with constant shaking (10). C2-depleted serum should be used immediately.

2. In the wells of a 96-well plate, mix 50 μL 2% ShEA in VBS, 50 μL C2-depleted serum (diluted 1/5 in VBS) and 50 μL of fraction to be tested for C2.
3. Incubate for 30 min at 37°C, spin and read the absorbance of the supernatant at 415 nm.
4. Fractions restoring C2 hemolytic activity are pooled for further use.

3.1.5. Purification of C3

C3: M_r 187 kDa; α: 112 kDa, β: 75 kDa; activated: C3a: 9 kDa, C3a': 102 kDa; serum conc. 1.3 mg/mL.

C3 is the most abundant C component in serum and can easily be purified in large quantities.

3.1.5.1. LARGE-SCALE PURIFICATION OF C3 *(11)*

1. To 1 L of plasma, add PMSF (0.5 m*M* final concentration) and PEG4000 to a final concentration of 5% (from a 15% stock in 100 m*M* NaPhosphate, 15 m*M* EDTA, 150 m*M* NaCl, 0.5 m*M* PMSF, pH 7.4); stir 30 min at 4°C.
2. Spin 30 min 10,000*g* at room temperature and make supernatant 12% PEG4000 by addition from a 26% stock in the same buffer, stir 30 min and spin 10,000*g* 30 min at 4°C.
3. Resuspend pellet in 100 mL of the same buffer and pass over a 100-mL lysine-Sepharose column to remove plasminogen.
4. Dilute breakthrough with 5 m*M* EDTA to a conductance of 3 mmho/cm (about 3× dilution), load onto a diethylaminoethyl (DEAE) cellulose column (5 cm × 40 cm) equilibrated in 25 m*M* NaPhosphate, 5 m*M* EDTA, pH 7.0 and elute with a 2-L linear NaCl gradient from 25 m*M* to 300 m*M*. Pool C3-positive fractions.
5. To the C3 pool, add PEG4000 to 16%, stir 30 min, and spin 30 min 10,000*g*, 4°C. Resuspend pellet in 15 mL 100 m*M* NaPhosphate, 5 m*M* EDTA, 50 m*M* aminocaproic acid (EACA), 0.5 m*M* PMSF, 150 m*M* NaCl, pH 7.4, and gel filter on a Sepharose 6B column (2.5 cm × 100 cm).
6. Dialyze against 25 m*M* KPhosphate, 100 m*M* KCl, 50 m*M* EACA (ε amino caproic acid), pH 7.4, load on a Hydroxylapatite column (2 cm × 20 cm) equilibrated in the same buffer, wash with the same buffer and elute first with 25 m*M* KPhosphate, 2 *M* KCl (elutes C5) followed by 125 m*M* KPhosphate, 100 m*M* KCl (elutes C3).
7. Pool and concentrate C3-containing fractions.

3.1.5.2. SMALL SCALE PURIFICATION OF C3 *(12)*

1. Dilute NHS (1–5 mL) 1/10 in PBS/5 m*M* EDTA/4% PEG6000 and spin 30 min 10,000*g*. Make supernatant 10% with PEG6000 by addition from a 30% PEG6000 stock solution in PBS/EDTA. Spin 30 min at 10,000*g*, 4°C.
2. Dissolve pellet to original volume in 20 m*M* Tris, 5 m*M* EDTA, pH 8.9. Apply to MonoQ column (0.5 mL/run) and elute with a 20-mL linear NaCl gradient from 0 to 500 m*M*.

3. Pool C3-positive fractions and gel filter on a Superose 12 FPLC column equilibrated in PBS to obtain pure C3.
4. Store in aliquots at −70°C.

3.1.5.3. GENERATION OF C3$_{MA}$ FOR COFACTOR ASSAYS

C3 with an inactivated thioester may be required for use in cofactor assays (*see* Chapter 18).

1. To C3 containing fractions off MonoQ column, add 1/10 (v/v) 2 *M* methylamine pH 8.0. Incubate 2 h 37°C.
2. Dialyze against MonoQ start buffer, apply to MonoQ and elute as above. Methylamine-reacted C3 (C3$_{MA}$) elutes a few fractions before native C3 from MonoQ *(13)*.

3.1.5.4. FUNCTIONAL ASSAY FOR C3

1. Generate C3/C4-depleted serum (R3) by incubation of fresh serum with 1/10 (v/v) 2 *M* methylamine pH 8.0 for 2 h at 37°C *(14)*.
2. Dialyze serum overnight against PBS and store in aliquots at −70°C.
3. In the wells of a 96-well plate mix 50 µL RaE (2% in VBS), zymosan (1 mg/mL final), 50 µL C3/C4-depleted serum diluted 1:10 in VBS and 50 µL of the test fraction.
4. Incubate for 30 min at 37°C, spin, and read the absorbance of the supernatant at 415 nm.
5. Fractions restoring C3 hemolytic activity are pooled for further use.

3.1.5.5. GENERATION OF FRAGMENTS

Generation of fragments of C3 can either be achieved using the C3 convertases of the C system or by enzymatic digestion using non-C enzymes.

3.1.5.5.1. Generation of C3a and C3b Using Cobra Venom Factor (CVF)-Bb

M_r; C3a: 9 kDa, C3b: 178 kDa *(15)*

1. Generate CVF-Sepharose solid phase by coupling CVF to CNBr-activated Sepharose (1 mg CVF/mL column volume) according to the manufacturer's instructions (for purification of CVF, *see* **Subheading 3.2.4.**). Wash solid phase and incubate with serum (30% suspension of CVF-Sepharose in neat serum) for 30 min at 37°C with constant mixing. A CVF-Bb complex will now be formed on the solid phase.
2. Wash CVF-Bb sepharose in PBS; add purified C3 (2 mg/mL; 1 mL per mL packed solid phase) and incubate 4 h at 37°C with constant stirring. Analyze the degree of conversion by sodium dodecyl sulfate-polyacrylamide gel electrophoresis (SDS-PAGE).
3. Separate C3a and C3b in the supernatant by gel filtration on Sephadex G-100 (2 cm × 100 cm), equilibrated in 20 m*M* NaPhosphate, 1 *M* NaCl, pH 7.0. C3b elutes in the breakthrough. Unreacted C3 can be separated from C3b by application on MonoQ, as described for C3 (*see* **Subheading 3.1.5.2.**).

3.1.5.5.2. Generation of iC3b and C3f:

M_r iC3b: 176 kDa, C3f: 2 kDa *(16)*.

1. Incubate C3b or $C3_{MA}$ (approx 1 mg/mL; generated as described in **Subheading 3.1.5.3.**) with factor I and factor H, both at 2% w/w of C3b/$C3_{MA}$, for 2 h at 37°C. Confirm digestion by running an aliquot on SDS-PAGE.
2. Fractionate mixture by gel filtration on a Sephadex G-100 column as described in **Subheading 3.1.5.1.** iC3b elutes in the breakthrough. For complete removal of fH and fI, apply the iC3b peak on MonoQ as described for C3.

3.1.5.5.3. Generation of C3c and C3dg

M_r C3c: 138 kDa, C3dg: 38 kDa *(17)*.

1. Incubate C3b or $C3_{MA}$ (approx 1 mg/mL; generated as earlier described) with factor I and CR1, both at 2% w/w of C3, at 37°C for 2 h. If purified CR1 or recombinant soluble CR1 are not available, a human erythrocyte ghost membrane extract can be used as a source of CR1.
2. Determine incubation time necessary to get complete cleavage by removing aliquots at intervals and running on SDS-PAGE.
3. Separate C3c and C3dg by gel filtration on Sephadex G-150 superfine in 10 mM NaPhosphate, 1 M NaCl, pH 7.4.

3.1.5.5.4. Generation of C3d and C3g

M_r C3d: 34 kDa, C3g: 4 kDa *(18)*.

1. Incubate C3dg (approx 1 mg/mL; generated as described in **Subheading 3.1.5.5.3.**) with trypsin (1% w/w) for 5 min at 37°C.
2. Stop reaction by adding 1 mM PMSF or 5% (w/w) solid-phase soybean trypsin inhibitor (SBTI).
3. Separate by gel filtration on Sephadex G-75 superfine in 10 mM NaPhosphate, 1 M NaCl, pH 7.4.

3.1.5.5.5. Generation of C3b, C3g, C3c, and C3d Using Trypsin *(18)*

1. For the simultaneous generation of C3g, C3c, and C3d, incubate C3 (approx 1 mg/mL) with trypsin (1% w/w) for 5 min at 37°C. For the tryptic generation of C3b, incubate C3 with 0.05–0.1% w/w trypsin for 2–3 min at 37°C. Fragments can also be generated using thrombin or plasmin at 2% w/w.
2. Stop with 1 mM PMSF or 5% (w/w) solid-phase SBTI.
3. Separate by gel filtration on Sephadex G-100 superfine in PBS.

3.2. Purification of Alternative Pathway Components

3.2.1. Purification of Factor B (fB)

M_r: fB: 93 kDa; Ba: 30 kDa, Bb: 63 kDa; serum conc.200–400 µg/mL *(19)*.

1. To 200 mL serum at 4°C, add PMSF (0.5 mM final), pepstatin A (2 µM), leupeptin (3 µM) and iodoacetamide (2 µM).
2. Slowly add 58 g $(NH_4)_2SO_4$ (50% saturation) and stir 1 h at 4°C, centrifuge 10,000g, 30 min, 4°C.
3. Filter supernatant through Whatman no. 1 filter paper and dialyze against 10 vol of 25 mM Tris, 0.5 mM $CaCl_2$, 0.5 mM $MgCl_2$ pH 7.4.
4. Add sodium caprylate (binds to albumin and prevents binding of albumin to column) to a final concentration of 25 mM and mix for 1 h at 4°C with 70 mL Cibacron Blue F3GA agarose equilibrated in the same buffer.
5. Pack the agarose in a 2 cm × 30 cm column, wash in the same buffer and elute with a 100 mL linear KCl gradient from 0 to 2 M KCl in the same buffer.
6. Identify fB-containing fractions, pool and dialyze against 10 mM KPhosphate, 5 mM EDTA including above-described inhibitors, pH 7.0 (start buffer).
7. Load onto a Hiload S-Sepharose column (1.6 cm × 10 cm) equilibrated in start buffer and elute with 100 mL linear NaCl gradient from 0 to 400 mM NaCl in start buffer.
8. Identify and pool the fB-containing fractions and reapply to the reequilibrated Hiload S-Sepharose column using the same conditions.
9. Identify, pool, dialyze and concentrate the fB-containing fractions, and store in aliquots at –20°C.

3.2.1.1. FUNCTIONAL ASSAY FOR FB

1. Generate fB-depleted serum by incubating 1–2 mL serum in a glass tube under continuous shaking at exactly 50°C for 20 min. This procedure also inactivates C2 and partially inactivates C6 and C7.
2. In the wells of a 96-well plate mix 2% RaE in APB (50 µL) with fB-depleted serum (1:5 in APB, 50 µL) and fractions to be tested for fB (50 µL). Incubate for 30 min at 37°C, spin and read the absorbance of the supernatant at 415 nm.
3. Fractions restoring fB hemolytic activity are pooled for further use.

3.2.2. Purification of Factor D (fD)

M$_r$: fD: 24 kDa; serum conc. 1–2 µg/mL.

Plasma concentration of fD is very low and alternative sources have often been used, notably, urine from patients with renal tubular dysfunction or peritoneal dialysis fluid. The following method has been developed to purify factor D from plasma *(20)*.

1. Serum (2 L) is dialyzed overnight at 4°C against 40 L of 5 mM EDTA pH 5.4.
2. The euglobulin precipitate is removed by centrifugation at 10,000g for 30 min at 4°C. To the supernatant, add 400 mL 0.5 M NaPhosphate, pH 6.0 to bring conductivity to 12–14 ms.
3. Apply to a CM-Sephadex C-50 column (8 cm × 12 cm) equilibrated in 200 mM NaPhosphate, pH 6.0. Elute with a 1-L linear NaCl gradient from 0 to 2 M NaCl in 100 mM NaPhosphate, pH 6.0.

4. Identify fD-containing fractions, pool, add $(NH_4)_2SO_4$ to 50% saturation (291 g/L), stir 2 h, 4°C, and centrifuge 10,000g, 1 h, 4°C.
5. Make supernatant 70% saturated with $(NH_4)_2SO_4$ by adding a further 125 g/L, stir 2 h at RT and spin 10,000g, 1 h.
6. Dissolve the pellet in 40 mL 0.1 M Tris, 2 mM EDTA pH 8.0; gel filter on a Sephadex G-75 column (8 cm × 90 cm) equilibrated in the same buffer. Identify and pool fD-containing fractions.
7. Apply to CM-cellulose 32 column (1.5 cm × 20 cm) equilibrated in 230 mM NaAc, pH 5.2 and elute with a 250-mL linear NaCl gradient from 0 to 300 mM.
8. Pool fD-containing fractions, concentrate, dialyze against 10 mM Tris, 1 mM MgCl$_2$, 1 mM CaCl$_2$, 150 mM NaCl, apply to a concanavalin A-Sepharose column (1 cm × 12 cm) equilibrated in the same buffer and collect breakthrough. Low-molecular weight contaminants can also be removed by passage over heparin-Sepharose.
9. Pool, concentrate by ultrafiltration using a 10-kDa filter, and store in aliquots at −70°C.

3.2.2.1. FUNCTIONAL ASSAY FOR FD

1. Generate fD-depleted serum by gel filtration of small volumes of serum on Sephadex G-75 (fD is separated because of its small size; it, alone of the C components, is significantly retarded on the column).
2. The column void fractions, containing the bulk of the serum proteins minus fD, are pooled and stored in aliquots at −70°C.
3. In the wells of a 96-well plate, mix RaE (50 µL of a 2% suspension in APB) with 50 µL of fD depleted serum diluted 1:5 in APB and 50 µL of the fraction to be tested for fD activity. Incubate for 30 min at 37°C, spin, and read the absorbance of the supernatant at 415 nm.
4. Fractions restoring fD hemolytic activity are pooled for further use.

3.2.3. Purification of Properdin

M_r: properdin: 54 kDa, 5 µg/mL; *(21)*.

The serum concentration of properdin is low and it strongly associates with activated C3; it is thus important to prevent C3 activation during purification or properdin will be lost. Properdin is present in plasma in the form of oligomers, mainly dimers (P_2), trimers (P_3), and tertramers (P_4). Methods have been described for the separation of the different properdin oligomers in plasma based upon gel filtration or cation exchange *(22,23)*.

1. Plasma (1 L) is adjusted to pH 6.0 by addition of dilute HCl and 1 mM PMSF and dialyzed overnight at 4°C against 10 vol of 5 mM EDTA, pH 5.4.
2. The euglobulin precipitate is removed by centrifugation at 10,000g for 30 min, dissolved in 100 mL 20 mM NaPhosphate, 2 mM EDTA , pH 6.0 and applied to a CM-Sephadex C50 column (2.5 cm × 40 cm) equilibrated in the same buffer. Elute using a 2-L linear NaCl gradient from 0 to 1 M NaCl in 20 mM NaPhosphate, 2 mM EDTA, pH 6.0.

3. Identify properdin-containing fractions, pool, dialyze into 20 mM NaPhosphate, 10 mM NaCl, 2 mM EDTA, pH 6.0, apply to a CM-cellulose CM32 column (2.5 cm × 40 cm) and elute using a 2-L linear NaCl gradient from 0 to 1 M NaCl in 20 mM NaPhosphate, 2 mM EDTA, pH 6.0.

4. Identify properdin-containing fractions, concentrate and gel filter on a Superdex HR200 column (2 cm × 100 cm) equilibrated in 50 mM Tris, 2 mM EDTA, 500 mM NaCl pH 7.0.

5. Pool, concentrate, and store in aliquots at −70°C.

3.2.3.1. FUNCTIONAL ASSAY

The functional assay is based on the ability of Properdin to stabilize the C3 convertase. Numerous complex assays have been proposed, but a simple hemolytic assay is adequate for testing column fractions.

1. In wells of a 96-well plate, mix 2% RaE in APB (50 μL) with various dilutions of serum (1:20; 1:10; 1:5 in APB, 50 μL) and fractions to be tested for properdin (50 μL).

2. Incubate for 30 min at 37°C, spin, and read the absorbance of the supernatant at 415 nm.

3. Identify fractions enhancing AP activity, pool, and concentrate.

3.2.4. Purification of Cobra Venom Factor (CVF)

CVF: M$_r$: 149 kDa; α: 69 kDa; β: 49 kDa; γ: 32 kDa *(24)*.

CVF is a C3-like molecule found in the venom of several different species of cobra. The purification method described is based on that described by Vogel and Müller-Eberhard *(24)*.

See **Note 2**.

1. Freeze-dried cobra venom (1 g; harvested from the cobra *Naja naja kaouthia*; available from Sigma in some parts of the world) is dissolved in 5 mM NaPhosphate, pH 7.2 (start buffer) and applied to a DEAE Sepharose fastflow column (5 cm × 14 cm) equilibrated in the same buffer.

2. After washing, the column is eluted with a 600-mL linear NaCl gradient from 0 to 400 mM in start buffer. CVF-containing fractions are identified, pooled, and dialyzed against 10 mM NaPhosphate, pH 5.0.

3. The dialyzed pool is applied to a CM-Sepharose fastflow column (2.6 cm × 30 cm) equilibrated in the same buffer and eluted with a 500-mL linear NaCl gradient from 0 to 600 mM in start buffer. Identify and pool CVF-containing fractions. Dialyze against 10 mM NaPhosphate, pH 5.0. Apply the CVF pool to an Affigel blue column (1.5 cm × 13 cm) equilibrated in the same buffer, wash in the same buffer, and elute bound CVF in 10 mM NaPhosphate, pH 6.5. Residual CVF and the contaminating PLA2 can be eluted from the Affigel column with 20 mM $(NH_4)_2SO_4$, pH 10.5.

4. Pool CVF-containing fractions, dialyze against VBS, concentrate to approx 1 mg/mL in an Amicon ultrafiltration cell using a 30-kDa cutoff membrane and store in aliquots at −70°C. Thawed aliquots remain active for several weeks when stored at 4°C.

3.2.4.1. FUNCTIONAL ASSAY FOR CVF

CVF from *Naja naja kaouthia* can be detected in a reactive lysis assay utilizing the ability of CVF from this source to form a C3/C5 convertase and activate membrane attack complex (MAC) assembly on a target cell.

1. In the wells of a 96-well plate, mix GpE (50 µL of a 2% suspension in VBS) with serum (50 µL of a 1/10 dilution in VBS) and 50 µL of the test fraction (neat and diluted 1:10 and 1:100 in VBS). Incubate 30 min at 37°C.
2. Control for PLA_2 contamination by parallel testing of fractions for lysis of GpE in the absence of serum (PLA_2 can cause hemolysis independent of serum).
3. Spin and read the absorbance of the supernatant at 415 nm.
4. Fractions inducing serum-dependent hemolysis of target GpE are pooled for further use.

3.2.4.2. ALTERNATIVE FUNCTIONAL ASSAY FOR CVF

CVF from the cobra *Naja haje* does not form a C5 convertase and must be detected using an inhibition assay.

1. In the wells of a 96-well plate, mix 50 µL of the test fraction with 50 µL human serum (1/20 in VBS) and incubate for 30 min at 37°C.
2. Add 50 µL ShEA (2% in VBS) and incubate for a further 30 min at 37°C.
3. Spin and read the absorbance of the supernatant at 415 nm. Fractions containing CVF show less lysis because of consumption of C in the first incubation.

3.3. Purification of Terminal Pathway Components

3.3.1. Purification of C5

M_r: C5: 192 kDa, α: 118 kDa, β: 74 kDa; activated: C5a: 11 kDa; C3α': 107 kDa; serum conc. 75 µg/mL *(11)*.

C5 copurifies with C3 through the procedure described in **Subheading 3.1.5.**

C5 is separated from C3 by Hydroxyapatite chromatography (**step 6** in the large-scale isolation of C3, **Subheading 3.1.5.**).

1. Identify C5-containing fractions off Hydroxyapaptite column from C3 preparation. Pool, dialyze against 20 m*M* Tris, 1 m*M* EDTA, 2.5 m*M* NaCl, pH 8.0, and apply to Q-Sepharose column (2 cm × 20 cm).
2. Wash with the above buffer and elute with a 500-mL linear NaCl gradient from 0 to 0.5 *M* in the same buffer.

3. Identify positive fractions, pool, concentrate, and gel filter on a Superdex HR200 column (2 cm × 90 cm) equlibrated and run in PBS.
4. Identify and pool positive fractions, concentrate and store at –70°C.

3.3.1.1. FUNCTIONAL ASSAY FOR C5:

C5 can be measured in a functional assay using C5-deficient mouse serum as a source of other C components. C5-deficient mice are readily available (*see* Chapter 20 for C5-deficient mouse strains) and the serum can be obtained commercially.

1. In the wells of a 96-well plate, mix antibody-sensitized rabbit E (50 μL of 2% in VBS) with 50 μL C5-deficient mouse serum (1/5 in VBS) and 50 μL of the test fraction.
2. Incubate 60 min at 37°C, spin, and read the absorbance of the supernatant at 415 nm.
3. Pool C5-containing fractions for further stages.

3.3.2. Purification of C6

M_r: C6: 108 kDa, serum conc. 70 μg/mL *(26)*.

1. To 2 L of serum at 4°C, add ammonium sulfate (213 g/L; 37.5% saturation), 1 m*M* PMSF and 1 m*M* benzamidine; stir 1 h at 4°C, and spin 10,000*g* for 30 min at 4°C.
2. Retain pellet for purification of C7 (**Subheading 3.3.3.**). Add ammonium sulfate to the supernatant (80 g/L; final saturation 50%), stir 1 h 4°C, and spin 30 min 10,000*g* at 4°C.
3. Dissolve precipitate in 100 mL 10 m*M* NaPhosphate, 1 m*M* PMSF, 1 m*M* Benzamidine, 2 m*M* EDTA, pH 6.2, and dialyze against the same buffer.
4. Apply to a phosphocellulose column (4 cm × 80 cm), wash and elute with a 2-L linear NaCl gradient from 0 to 1 *M*.
5. Identify C6-containing fractions, pool, concentrate, and dialyze into 10 m*M* NaPhosphate, 1 m*M* PMSF, 1 m*M* Benzamidine, 2 m*M* EDTA, 25% glycerol, pH 7.8.
6. Load on a Q-Sepharose column (2.5 cm × 80 cm) equilibrated in the same buffer, wash and elute with a 2-L linear NaCl gradient from 0 to 500 m*M*.
7. Identify and pool C6-containing fractions, concentrate and gel filter on a Sephadex G200 column (2 cm × 90 cm) equilibrated in VBS.
8. Identify and pool positive fractions, concentrate and store at –70°C.

3.3.2.1. FUNCTIONAL ASSAY FOR C6

C6 can be measured in a functional assay using C6-deficient rabbit or rat serum (*see* Chapter 20) or C6-depleted serum obtained by passage of NHS over an anti-C6 immunoaffinity column (Chapter 4).

1. In the wells of a 96-well plate, mix ShEA (50 μL of 2% in VBS) with 50 μL C6-deficient rat serum (1/5 in VBS) and 50 μL of the test fraction.

2. Incubate 60 min at 37°C, spin, and read the absorbance of the supernatant at 415 nm.
3. Pool C6-containing fractions for further stages.

3.3.3. Purification of C7

M_r: C7: 121 kDa; serum conc. 60 µg/mL *(26)*.

1. Resuspend the 37.5% ammonium sulfate pellet obtained in stage 1 of the C6 preparation (**Subheading 3.3.2.**) in 150 mL NaPhosphate, pH 6.0, and dialyze against the same buffer.
2. Apply to a Phosphocellulose column (4 cm × 80 cm) equilibrated in the same buffer, wash, and elute with a 2-L linear NaCl gradient from 0 to 1 *M* in the same buffer. Identify C7-containing fractions, pool, and dialyze against 10 m*M* NaPhosphate pH 7.0.
3. Apply to a QAE-Sephadex column (2.5 cm × 40 cm) equilibrated in 10 m*M* NaPhosphate, 25% glycerol, pH 7.0, wash, and elute with a 2-L linear NaCl gradient from 0 to 500 m*M*.
4. Identify and pool C7-containing fractions, concentrate and gel filter on a Sephadex G-200 column (2 cm × 90 cm) equilibrated in VBS.
5. Identify and pool positive fractions, concentrate and store at –70°C.

3.3.3.1. FUNCTIONAL ASSAY FOR C7

C7 can be measured in a functional assay using C7-deficient serum or C7-depleted serum obtained by passage over an anti-C7 immunoaffinity column (Chapter 4).

1. In the wells of a 96-well plate, mix antibody-sensitized sheep E (50 µL of 2% in VBS) with 50 µL C7-depleted serum (1/5 in VBS) and 50 µL of a dilution of the test fraction.
2. Incubate 60 min at 37°C, spin, and read the absorbance of the supernatant at 415 nm.
3. Pool C7-containing fractions for further stages.

3.3.4. Purification of C8

M_r: C8: 150 kDa; nonreduced: α-γ: 86 kDa β: 64 kDa; reduced: α: 64 kDa; β: 64 kDa; γ: 22 kDa. Serum conc. 80 µg/mL *(27)*.

1. To 0.5 L plasma or serum at 4°C, add PMSF and benzamidine both to 1 m*M* final concentration and $BaCl_2$ to 40 m*M*; mix 15 min at 4°C and centrifuge at 6000*g* for 15 min at 4°C.
2. To the supernatant at 4°C, add PEG4000 from a 20% stock solution to a final concentration of 5%, stir for 30 min, spin 6000*g* 15 min 4°C.
3. Make supernatant 10% PEG4000 by addition of solid PEG4000, stir 30 min, and spin 15 min 6000*g*, 4°C. Remove and retain supernatant for purification of C9 (**Subheading 3.3.5.**). Dissolve pellet in 25 m*M* imidazole, 150 m*M* NaCl, 1 m*M* benzamide, 1 m*M* PMSF pH 6.1, and remove insoluble material by centrifugation for 15 min at 6000*g*, 4°C.

4. Add solid $(NH_4)_2SO_4$ to 37.5% saturation (213 g/L) at 4°C. After 30 min stirring, spin at 6000g for 15 min at 4°C, discard pellet, and adjust supernatant to 50% saturation $(NH_4)_2SO_4$ (80 g/L), stir further for 1 h, and spin 6000g, 15 min 4°C. Discard supernatant, solubilize the pellet in 100 mL 70 mM NaPhosphate, 50 mM NaCl, 1 mM benzamidine HCl, 1 mM PMSF, pH 6.1 (CM buffer), and dialyze against the same buffer.
5. Apply to a CM-Sephacel column (5.5 cm × 25 cm) equilibrated in CM buffer. Wash and elute with a 600-mL linear NaCl gradient from 50 to 500 mM in CM buffer. Identify and pool positive fractions, concentrate, and dialyze against 25 mM Tris, 70 mM NaCl, 1 mM benzamidine, pH 8.0 (Q buffer).
6. Apply to a Q-Sepharose column (3 cm × 34 cm) equilibrated in Q buffer. Wash and elute with a 600-mL linear NaCl gradient from 70 to 220 mM in Q buffer. Identify and pool positive fractions and concentrate using a 30-kDa cutoff filter.
7. Apply to a Sephacryl S-300 gel filtration column (3.5 cm × 52 cm) equilibrated in 25 mM Imidazole, 150 mM NaCl, 0.02% NaAzide, pH 7.2. Identify and pool positive fractions, concentrate, add 20% (v/v) glycerol (to prevent aggregation which occurs on freezing) and store in aliquots at –20°C.

3.3.4.1. FUNCTIONAL ASSAY FOR C8

C8 can be measured in a functional assay using C8-deficient serum or C8-depleted serum obtained by passage over an anti-C8 immunoaffinity column (Chapter 4).

1. In the wells of a 96-well plate, mix ShEA (50 µL of 2% in VBS) with 50 µL C8-depleted serum (1/5 in VBS) and 50 µL of the test fraction.
2. Incubate 60 min at 37°C, spin, and read the absorbance of the supernatant at 415 nm.
3. Pool C8-containing fractions for further stages.

3.3.5. Purification of C9

M_r: C9: 66 kDa; serum conc. 60 µg/mL *(27)*.

1. To the 10% PEG4000 supernatant obtained in **Subheading 2.5.4. (step 3)**, add solid PEG4000 to a final concentration of 25%, stir 30 min, and spin 15 min 6000g, 4°C.
2. Discard supernatant, dissolve pellet in 10 mM NaPhosphate, 10 mM EDTA, 90 mM NaCl, 1 mM PMSF pH 7.4 (DE buffer), and centrifuge 15 min, 6000g, 4°C to remove insoluble material.
3. Apply supernatant to a DEAE-Sephacel column (5 cm × 10 cm) equilibrated in DE buffer, wash, and elute with a 1-L linear NaCl gradient from 90 mM to 500 mM in DE buffer. Identify and pool C9-containing fractions (*see* **Note 3**).
4. Apply directly to a Hydroxylapatite column (3.5 cm × 22 cm) equilibrated in 50 mM KPhosphate, 100 mM NaCl, pH 8.3, wash, and elute with a 1-L linear KPhosphate gradient from 50 mM to 300 mM.

5. Identify, pool, and concentrate C9-containing fractions and apply to a Sepharose CL-6B gel filtration column (1.5 cm × 46 cm) equilibrated and run in VBS. Identify, pool, and concentrate positive fractions and store at –70°C.

3.3.5.1. FUNCTIONAL ASSAY FOR C9

C9 can be measured in a functional assay using C9-depleted serum obtained by passage over an anti-C9 immunoaffinity column (Chapter 4).

1. In the wells of a 96-well plate, mix antibody-sensitized sheep E (50 μL of 2% in VBS) with 50 μL C9-depleted serum (1/5 in VBS) and 50 μL of a dilution of the test fraction.
2. Incubate 15 min at 37°C, spin, and read the absorbance of the supernatant at 415 nm.
3. Pool C9-containing fractions for further stages.

3.3.6. Purification of the C5b6 Complex

M_r: C5b6: 328 kDa *(2)* (*See* **Note 3**).

1. Incubate 1 L of C7-depleted serum with 10 mg/mL zymosan overnight at room temperature to generate C5b6.
2. Remove zymosan by centrifugation and dialyze serum against 20 mM NaPhosphate pH 5.4.
3. Spin 30 min at 10,000g and dissolve the euglobulin precipitate in 50 mL 10 mM NaPhosphate, 5 mM NaCl, 25% glycerol, pH 7.0 (DE buffer).
4. Apply to DEAE Sephacel column (5 × 30) equilibrated in DE buffer. Elute with 5-L linear gradient from 0 to 500 mM NaCl in DE buffer. Pool and concentrate C5b6-containing fractions and apply to Sephadex-G-200 (2 cm × 100 cm) equilibrated in 10 mM NaPhosphate, 500 mM NaCl, pH 7.0. Pool C5b6-containing fractions, concentrate and store at –70°C.

3.3.6.1. FUNCTIONAL ASSAY

Numerous functional assays have been used, all relying on the principle of reactive lysis. One simple protocol is to measure lysis of GpE in agarose.

1. Warm 5 mL of a 1.5% suspension of GpE in PBS/10 mM EDTA to 40°C and mix with 5 mL of a 2% solution of agarose (in PBS/10 mM EDTA) also at 40°C.
2. Immediately pour mixture into a Petri dish or onto an appropriate glass plate and allow to set.
3. Punch sets of holes in agarose, each set comprising a central hole surrounded by six holes all equidistant from center.
4. Place NHS (10 μL) in central hole and fractions to be tested (10 μL) in surrounding holes.
5. Incubate at RT for 4–8 h, look for lines of hemolysis (clearing) between central well and wells containing positive fractions.
6. Pool positive fractions and concentrate for next step.

3.4. Purification of the Anaphylatoxins: C3a, C4a, *and* C5a

1. To fresh serum, add the protease inhibitors 6-aminohexanoic acid (1 M) and either 2-mercaptomethyl-5-guanidinopentanoic acid (1 mM) or 2-mercaptomethyl-3-guanidinoethylthiopropranoic acid (1 mM) *(28)*.
2. Add zymosan at 20 mg/L to activate AP and generate C3a and C5a and, after 10 min at 37°C, add aggregated IgG at 0.5 mg/mL (prepare by heating IgG at 10 mg/mL in PBS at 63°C, 20 min) to activate CP and generate C4a. Continue incubation for a further 30 min at 37°C.
3. Add 10 M HCl to a final concentration of 1 M and cool on ice for 1 h.
4. Remove precipitate by centrifugation and dialyze supernatant against 100 mM Ammonium formate, pH 5.0 (use dialysis tubing with 3.5 kDa cutoff), and concentrate to 1/5 of original volume in an ultrafiltration cell (5 kDa cutoff).
5. Apply to Biogel P-60 gel-filtration column (2 cm × 100 cm) equilibrated in 100 mM ammonium formate, pH 5.0, identify positive fractions, pool and concentrate.
6. Apply to a SP-Sephadex column (1.6 × 30) equilibrated in the same buffer. Wash with 100 mM ammonium formate pH 7.0 to elute contaminating proteins. Elute bound protein with a 650 mL linear gradient to 800 mM ammonium formate pH 7.0. Anaphylatoxins elute in the order: C5a, C4a, C3a.
7. Apply C5a containing fractions to CM-Sephadex C-25 (0.6 cm × 14 cm) equilibrated in 0.1 M ammonium formate pH 7.0 and elute with equilibration buffer.
8. Dialyze C4a fractions against 150 mM ammonium formate pH 5.5 and apply to CM-cellulose (CM52; 1.6 × 15) equilibrated in the same buffer and elute with 400 mL linear gradient to 350 mM ammonium formate.
9. Dialyze C3a fractions against 150 mM ammonium formate pH 7.0 and apply to CM-cellulose (CM52; 1.5 × 27) equilibrated in the same buffer and elute with 600 mL linear gradient to 450 mM ammonium formate *(see* **Note 4**).

3.4.1. Functional Assays

Numerous assays for the activities of the anaphylatoxins have been used, including smooth muscle contraction, neutrophil chemotaxis assays, and neutrophil oxidase assays. Such methods are beyond the scope of this chapter. For identification of C3a and C5a in column fractions the simplest approach is to use specific antisera in immunochemical assays.

3.5. Purification of Complement Regulators

A variety of functional assays can be used to measure the inhibitory activity of the fluid phase and membrane bound regulators. A large number of regulators have cofactor activity for factor I (C4bp, fH, CR1, MCP). The method to measure cofactor activity is fairly general for all inhibitors and is described later and in more detail in Chapter 18. Some membrane bound regulators (DAF, CD59) have the ability to reincorporate into a membrane and so confer C resistance to the target cell, however, all these inhibitors require a different target cell and the methods will be described with the respective purification methods. Fluid phase regulators are generally purified from serum. For the membrane bound regulators and receptors, it is important to choose the right cell type. Erythrocytes are the most convenient source of C regulators because they can easily be obtained in large quantities, are enucleate, and do not contain large amounts of proteolytic enzymes. Other cell types that can be obtained in relatively large quantities are platelets, neutrophils, tonsils, and spleen cells. An alternative is to use a cell line expressing the regulator or receptor of interest, but this will usually require a large volume of cultured cells. Membrane bound C regulatory proteins are either attached to the membrane via a transmembrane anchor (TM), like CR1 and MCP, or a glycosyl phosphatidylinositol (GPI) anchor (DAF, CD59). The first step in purification of a C regulator is membrane extraction, to be described in **Subheading 3.5.7.** Erythrocytes are the best source of DAF, CD59 and CR1 (MCP is not expressed on human erythrocytes), whereas platelets or an appropriate cell line can be used for purification of MCP.

3.5.1. Purification of C1-inh

C1-inh: M_r 110 kDa, serum conc. 200 µg/mL *(29)*.
Small scale purification of C1-inh can be carried out as follows:

1. Dilute 5 mL serum to 50 mL with 15% PEG6000 in PBS/5 mM EDTA, spin 30 min 10,000g, 4°C, and retain supernatant.
2. Increase PEG concentration in the supernatant to 25% by addition of solid PEG6000. Stir 30 min and spin 10,000g, 4°C, 30 min.
3. Dissolve precipitate in 5 mL 20 mM Tris, pH 7.0, and apply to a Heparin Sepharose column (1.6 × 8) equilibrated in the same buffer. Elute with 60 mL of a linear gradient from 0 to 1 M NaCl.
4. Pool C1-inh-containing fractions, dilute four times in 20 mM Tris, pH 7.0 and apply to a MonoQ column equilibrated in the same buffer. Elute with a 20-mL linear gradient to 500 mM NaCl.
5. Identify C1-inh-containing fractions, pool, concentrate and fractionate on Superose 12 in PBS. Concentrate C1-inh pool and store in aliquots at –70°C.

3.5.1.1. FUNCTIONAL ASSAY FOR C1-INH

The functional assay of C-inh is based on the inhibition of C1s esterolytic activity (*see* **Subheading 3.1.1.**).

1. In wells of a 96-well plate, mix C1-inh containing fractions with C1s (1 µg/well) in a total volume of 100 µL and incubate for 30 min at 37°C.
2. To each well, add chromogenic substrate specific for C1s (*see* **Subheading 3.1.1.**)
3. Incubate 30 min 37°C and measure absorbance of the cleaved chromogenic substrate.
4. Pool fractions that inhibit C1s esterolytic activity.

3.5.2. Purification of C4 Binding Protein (C4bp)

C4bp: M_r 450–550 kDa, reduced: 72 kDa. 6–7 α chains: 70 kDa, 0–1 β chain: 45 kDa, serum conc. 250 µg/mL *(30)*.

1. Collect blood in 10 times concentrated buffered solution to obtain final conc. of 5 mM NaPhosphate, 10 mM benzamidine, 10 mM EDTA and 0.1 mM PMSF, pH 7.4, and separate cells from plasma by centrifugation.
2. To 500 mL buffered plasma, add 250 mL 15% (w/v) PEG4000 (in 5 mM NaPhosphate, 10 mM EDTA, 0.1 mM PMSF, 75 mM NaCl, pH 7.4) to achieve a final PEG4000 concentration of 5%, centrifuge at 10,000g 30 min.
3. Resuspend precipitate in 20 mL 50 mM NaPhosphate, 1 M NaCl, 5 mM EDTA, 5 mM EACA, 0.5 mM PMSF, pH 6.7 (GF buffer).
4. Apply to Sepharose 6B gelfiltration column (5 cm × 50 cm) equilibrated and run in GF buffer. Identify and pool C4bp-containing fractions.
5. To pooled fractions add 0.5 vol of 22.5% (w/v) PEG4000 in 50 mM Tris, 75 mM NaCl, 0.5 mM CaCl$_2$, pH 8.0, and stir 30 min. Centrifuge 10,000g 30 min and resuspend pellet in 80 mL of 50 mM Tris, 75 mM NaCl, 0.5 mM CaCl$_2$, 0.5 mM PMSF, pH 8.0 (HS buffer).
6. Apply to heparin Sepharose column (2.5 × 70) equilibrated in HS buffer. Elute with 1-L linear gradient 0 to 0.6 M NaCl in HS buffer.
7. Add 0.5 vol of 22.5% (w/v) PEG4000 in 50 mM Tris, 20 mM NaCl, 0.5 mM CaCl$_2$, 0.5 mM PMSF, pH 8.0. Centrifuge 10,000g 30 min and resuspend pellet in 70 mL 50 mM Tris, 20 mM NaCl, 0.5 mM CaCl$_2$ (DE buffer).
8. Apply to DE-52 anion exchange column (2.5 cm × 30 cm) equilibrated in DE buffer and elute with 500-mL linear gradient 0 to 0.4 M NaCl in DE buffer.
9. Concentrate by ultrafiltration and apply to an affinity column consisting of C4 coupled to Sepharose (5 mg C4 coupled to 5 mL CNBr-activated sepharose) equilibrated in 10 mM Tris, 15 mM NaCl, 0.1 mM CaCl$_2$, pH 8.1.
10. Elute bound C4bp in 50 mM NaAc, 300 mM NaCl, pH 6.5, dialyze into VBS, concentrate and store at −70°C.

3.5.2.1. FUNCTIONAL ASSAY FOR C4BP

Functional assay is based on the capacity of C4bp to act as cofactor for factor I in the cleavage of iC3b. In the assay, methylamine treated human C3 (C3$_{MA}$) is used as a substrate.

1. Dialyze fractions containing cofactor (factor H, CR1, or MCP) into PBS/0.1% NP40.
2. Mix with 200 µg methylamine treated human C3 (*see* **Subheading 3.1.5.**) and 20 µg human factor I and incubate 2 h or overnight at 37°C.
3. Run on 10% SDS-PAGE under reducing conditions, blot onto nitrocellulose and probe blots with anti-C3. Cofactor activity is revealed by the disappearance of the α-chain and the appearance of the 43 and 47 kDa α-fragments.
4. To measure cofactor activity of C4bp, methylamine treated human C4 (C4$_{MA}$) is used as a substrate, but the assay is otherwise similar.

3.5.3. Purification of Factor I

fI: 2 chains: M_r 50 Kda, 38 kDa, serum concentration 30–50 µg/mL *(31)*.

1. Add 1 mM PMSF to 500 mL of fresh plasma and pass over a lysine Sepharose column (5 cm × 15 cm) equilibrated in 100 mM KPhosphate, 150 mM NaCl, 15 mM EDTA, pH 7.0. Collect breakthrough.
2. Dialyze against 20 mM Tris, 60 mM NaCl, 1 mM PMSF, pH 7.8, and load onto QAE-Sephadex column (7 cm × 36 cm) equilibrated in the same buffer. Elute with a 4-L linear gradient of 60–360 mM NaCl.
3. Pool and make 60% saturated with $(NH_4)_2SO_4$ (36 g/100 mL), stir 2 h, 4°C, and spin 20 min, 8000g, 4°C.
4. Dissolve precipitate in 60 mL 50 mM NaPhosphate, 200 mM NaCl, 1 mM PMSF, pH 7 and pass over a wheat germ lectin column (Pharmacia; 4 cm × 7.5 cm) equilibrated in the same buffer. Elute with buffer containing 100 mg/mL N-acetyl-D-glucosamine.
5. Dialyze against 25 mM KPhosphate, 1 mM PMSF, pH 7.5, and load onto Hydroxylapatite column (3.2 cm × 15 cm) equilibrated in the same buffer. Elute with 600 mL gradient to 200 mM KPhosphate
6. Concentrate by addition of 40 g/mL $(NH_4)_2SO_4$, stir 2 h, 4°C, and spin 20 min, 8000g, 4°C.
7. Dissolve precipitate in 5 mL 25 mM Tris, 150 mM NaCl, pH 7.0, and pass over Sephacryl-200 gel filtration column (2 cm × 90 cm) equilibrated in the same buffer. Store at 4°C

3.5.3.1. Functional Assay for fI

Cofactor assay using $C3_{MA}$ and appropriate cofactor (e.g., erythrocyte cell extract as source of CR1 or purified MCP, CR1, or fH) essentially as described in **Subheading 3.5.2.**

3.5.4. Purification of Factor H

FH: M_r 170 kDa, serum concentration: 200–600 µg/mL; 20 SCR *(32)*.

1. Bring 1 L of serum to 5% PEG4000 by addition of 250 mL of 25% (w/v) PEG4000 in 50 mM NaPhosphate, 150 mM NaCl, 15 mM EDTA, pH 7.4, stir for 30 min, and spin 30 min 8000g.
2. Bring supernatant to 12.5% PEG4000 by addition of 190 mL of 50% (w/v) PEG4000 in the same buffer, stir 30 min, and spin 8000g, 30 min. Dissolve precipitate in 300 mL 50 mM NaPhosphate, 15 mM EDTA, 150 mM NaCl pH 7.0.
3. Pass over lysine-Sepharose column (5 cm × 15 cm) and dialyze against 25 mM KPhosphate, 5 mM EDTA, pH 7.0 (DE buffer).
4. Load on DEAE-Sephadex column (5 cm × 50 cm) equilibrated in DE buffer and elute with 4 L of a linear gradient 0 to 500 mM NaCl in DE buffer.
5. Pool fI-containing fractions and bring to 14% (w/v final conc.) PEG4000 by addition of appropriate volume of 50% stock solution, stir 30 min, and spin 8000g, 30 min.

Resuspend precipitate in 20 mL 100 mM KPhosphate, 5 mM EDTA, 150 mM EDTA, pH 7.4.

6. Apply to Sepharose 6B gelfiltration column (5 cm × 90 cm) equilibrated in the same buffer and dialyze against 10 mM KPhosphate, pH 6.8 (HA buffer).
7. Apply to Hydroxyapatite column (1.5 cm × 20 cm) equilibrated in HA buffer. Elute with 500 mL of a linear gradient 0 to 200 mM KPhosph in HA buffer. Dialyze against 10 mM Tris, 1 mM EDTA, 25 mM NaCl, pH 8.4 (DE buffer).
8. Apply to DEAE-Sepharose CL-6B column (2 cm × 25 cm) equilibrated in DE buffer and elute with 1.4 L linear gradient 25–400 mM NaCl in DE buffer.

3.5.4.1. FUNCTIONAL ASSAY FOR FACTOR H

Cofactor assay essentially as described in **Subheading 3.5.2.**

3.5.5. Purification of S-protein

S-protein: M_r:89 kDa, reduced 84 kDa and 69 and 15 kDa fragments. Serum concentration 0.35–0.5 mg/mL. Alternative name: vitronectin *(33)*.

1. To 1 L of fresh citrated plasma add the following inhibitors: 10 mM benzamidine, 1 mM PMSF, 0.5 mM NPGB. Add slowly while stirring, 80 mL of 1 M BaCl$_2$, stir for 1 h, and spin 15 min 6000g 4°C.
2. To the supernatant, add PEG4000 (from a 50% stock solution in 10 mM Tris, 150 mM NaCl, pH 7.4) to a final concentration of 9%. Stir for 1 h and spin 6000g, 15 min, 4°C.
3. To the supernatant, add solid PEG4000 to give a final concentration of 20%. Stir for 1 h and spin 6000g, 15 min 4°C. Dissolve pellet in 20 mM NaPhosphate, 2 mM EDTA, 2 mM benzamidine, 1 mM reduced glutathione, 1 mM PMSF, pH 7.0 (DE buffer), and spin at 6000g, 15 min 4°C.
4. Load onto DEAE-Sephacel column (2.9 cm × 40 cm) equilibrated in DE buffer. Wash with DE buffer containing 25 mM NaCl, elute with a 2-L linear gradient from 25 to 300 mM NaCl. S-protein coelutes with ceruloplasmin (blue fractions off column) and just before C9, S-protein eluting after C9 is usually aggregated.
5. Pool positive fractions and add solid (NH$_4$)$_2$SO$_4$ to give 70% final saturation. Stir for 1 h and spin 10,000g, 20 min 4°C. Dissolve precipitate in 50 mM Tris, 150 mM NaCl, 2 mM benzamidine, 1 mM glutathione pH 7.4, and dialyze against the same buffer.
6. Apply to Blue-Sepharose column (2.5 cm × 25 cm) equilibrated in the same buffer and elute with 1 M NaCl.
7. Pool positive fractions and concentrate by ultrafiltration using a YM10 membrane.
8. Apply to a Sephacryl S-200 gel filtration column (2.8 cm × 105 cm) equilibrated in 50 mM Tris 150 mM NaCl, 2 mM benzamidine, 1 mM glutathione pH 7.4.
9. Pool positive fractions and apply to an antihuman albumin immunoaffinity column equilibrated in the same buffer, collect breakthrough.
10. Apply to an anti-C9 immunoaffinity column and collect breakthrough.
11. Concentrate to 1 mg/mL and store in small aliquots at −70°C (*see* **Note 5**).

3.5.6. Purification of Clusterin

Clusterin: M_r approx 80 kDa nonreduced, two chains (α and β) at 35–40 kDa reduced, 0.2–0.4 mg/mL in serum *(34)*. Methods for clusterin purification all involve the use of immunoaffinity chromatography, which is described in Chapter 4.

3.5.7. General Procedures for Membrane Extraction

3.5.7.1. ERYTHROCYTE GHOST PREPARATION

1. Wash 4 U (approx 1 L) of packed erythrocytes three times with PBS to remove plasma and "buffy coat" containing white cells.
2. Lyse packed cells in 10 vol of ice-cold 5 mM NaPhosphate, 2 mM EDTA. Additional protease inhibitors (1 mM benzamidine, 1 mM PMSF, 0.02% sodium azide) are required for some proteins. Stir overnight at 4°C (if the lysis mix warms up, ghosts will reseal and include hemoglobin, resulting in less clean ghosts.)
3. Wash and concentrate ghosts using a pellicon ultrafiltration system (membrane 300 kDa cutoff) and/or spin 13,000g, 30 min, 4°C until supernatant is almost free of hemoglobin and ghost pellet appears faintly pink. Note that the ghost pellet is very easily disturbed and great care must be taken in removing wash buffer to avoid loss of ghosts.

3.5.7.2. MEMBRANE EXTRACTION

GPI-anchored DAF and CD59 can be extracted using N-butanol or 1% NP40. Butanol extraction yields cleaner preparations and is best for subsequent classical chromatography purification. Butanol extraction is not appropriate for TM anchored CRP or CR. Extraction with NP40 is simpler, can be used for TM proteins and gives better yields making it most appropriate for subsequent affinity chromatography.

3.5.7.2.1. Butanol Extraction

1. To 1 part of packed ghosts, add three parts of PBS and one part of N-butanol. Stir for 30 min at RT, and spin 10 min, 2500g.
2. Remove the lower aqueous phase and dialyze against the starting buffer for the first purification step. Change dialysis buffer several times to ensure removal of all N-butanol.

3.5.7.2.2. NP40 Extraction

1. Take one part of ghosts, add four parts of PBS and add NP40 to 1% final concentration. Stir 30 min at RT.
2. Spin to remove undissolved material 15 min, 10,000g and use supernatant for further purification.

3.5.7.3. MEMBRANE EXTRACTION OF PLATELETS OR CELL LINES

For extraction of membrane bound proteins from nucleated cells it is necessary to add protease inhibitors to prevent degradation of the protein of interest.

1. Wash cells (platelets, leukocytes, or appropriate cell line) three times in PBS to remove serum proteins.
2. Lyse cells in ice cold lysis buffer (20 mM Tris pH 8.2, 140 mM NaCl, 2 mM EDTA, 1% NP40, 2 μg/mL aprotinin, 2 μg/mL leupeptin, 1 mM PMSF) at 1–5 × 10^{7}cells/mL for 30 min on ice. To increase yield, other detergents like digitonin can be used, however, these are much more expensive than NP40.
3. Spin to remove undissolved material 15 min 10,000g and use the supernatant for further purification.

3.5.8. Purification of Membrane Cofactor Protein (MCP/CD46)

MCP: M$_r$ 50–80, single-chain, TM anchor, 4 SCR *(35)*.

A major problem with the purification of MCP is its tendency to stick nonspecifically to various surfaces, including column matrices and filters. A standard chromatography method to purify MCP is carried out as follows.

1. Dialyze a 1% NP40 extract from 20 mL of packed cells (~ 3 × 10^{10} cells, e.g., U937 or platelets) against 20 mM Na-acetate, 0.1% NP40, pH 5.0. Remove precipitate by centrifugation 15 min 10,000g.
2. Apply to a chromatofocusing column (2 cm × 50 cm) equilibrated in the same buffer, pH 4.75. Elute the column with a linear gradient (20–250 mM) acetate buffer (pH 4.3) in 10% (vol/vol) polybuffer 74. Pool and dialyze positive fractions against 10 mM Tris, 0.05% NP40, pH 7.5.
3. Apply to a 50-mL high-resolution Hydroxyapatite column, equilibrated in the same buffer. Elute the column with a linear KPhosphate gradient from 0 to 250 mM pH 7.5 containing 0.05% NP40. Pool positive fractions.
4. Load onto a C3$_{MA}$-sepharose column (20 mL sepharose to which 20 mg C3MA (generated and purified as described in **Subheading 3.1.5.**) has been immobilized, equilibrated in 0.05% NP40, pH 7.2. Wash with 20 mM NaPhosphate 1 mM PMSF, 0.05% NP40, pH 7.0. Elute with the same buffer containing 0.5 M NaCl. Dialyze against the same buffer containing 20 mM NaCl.
5. Load onto a MonoQ column in equilibrated in the same buffer and elute with a linear NaCl gradient from 20 to 500 mM.

3.5.8.1. FUNCTIONAL ASSAY

Cofactor assay essentially as described in **Subheading 3.5.2.**

3.5.9. Purification of Decay Accelerating Factor (DAF/CD55)

DAF: M$_r$ 70 kDa, single-chain, GPI anchored, 4 SCR *(36)*.

1. Dialyze N-butanol extract of erythrocytes into 20 mM Tris, 40 mM NaCl, pH 8.0 (DE buffer), and apply to a DEAE Sepharose column (5 cm × 30 cm) equilibrated in DE buffer. Wash in DE buffer containing 0.1% NP40 and elute with a linear

gradient 40–500 mM NaCl in DE buffer with NP40. Concentrate active fractions and dialyze against 20 mM NaPhosphate, 0.1% NP40, pH 7.4 (HA buffer).

2. Apply to Hydroxyapatite column (5 cm × 20 cm) equilibrated in HA buffer. Wash and elute with 1 L linear Na/KPhosphate gradient in HA buffer. Pool active fractions.

3. Add NaCl to 300 mM and apply to phenyl-Sepharose (1.5 × 6.5) column equilibrated in 40 mM NaPhosphate, 300 mM NaCl, 0.1% NP40, pH 7.4. Elute with 160 mL inverse linear gradient to from 300 to 40 mM NaPhosphate in 50 mM NaCl, 1% NP40, pH 9.5.

4. Dialyze pooled active fractions against 5 mM NaPhosphate, 25 mM NaCl 0.1% NP40, pH 7.5, apply to a trypan blue-Sepharose column (2 cm × 5 cm) and elute with a 100-mL NaCl gradient 25 to 325 mM in the same buffer.

5. Identify active fractions, pool, dialyze against PBS-CHAPS (0.05%), concentrate by ultrafiltration and store at 4°C with 0.01% NaN3 or in small aliquots at –20°C.

3.5.9.1. FUNCTIONAL ASSAY

DAF has the capacity to reincorporate into a membrane through its GPI anchor and so confer resistance to C. The ideal target cells for measuring DAF activity are sheep erythrocytes. No interference of contaminating CD59 will be observed because of the high level of expression of functional sheep CD59 that masks the activity of any addition human CD59 added *(37,38)*.

1. Incubate column fractions with 2% ShEA for 30 min at 37°C (NB fractions have to be dialyzed into 0.05% CHAPS before assay as NP40 will lyse the cells).

2. Wash cells three times in VBS and resuspend in VBS at 2%.

3. In the wells of a 96-well plate, incubate with dilutions of human serum (1/50–1/100) for 30 min at 37°C.

4. Measure absorbance at 412 nm in supernatant and plot percent lysis at each serum dose (*see* **Note 5**).

3.5.10. Purification of CD59

CD59: M$_r$ 18–20 kDa, single-chain, GPI-anchored *(37)*.

3.5.10.1. PURIFICATION BY ANION EXCHANGE AND HYDROPHOBIC INTERACTION:

1. Dialyze butanol extract derived from 1 U (250 mL) of human erythrocytes against 20 mM Tris, 0.05% CHAPS, pH 8.4 (Q buffer), and apply to a Q-Sepharose fast-flow column (1.6 cm × 15 cm) equilibrated in the Q buffer. Elute with 600 mL of a linear gradient to 0 to 1 M NaCl. Pool positive fractions and dialyze against 20 mM NaPhosphate, 0.05% CHAPS, pH 7.4 (Q buffer).

2. Apply to MonoQ column equilibrated in Q buffer. Elute with 30 mL of a linear gradient to 0.5 M NaCl. Bring to 1.7 M by addition of solid ammonium sulphate.

3. Apply to phenyl superose column equilibrated with 40 mM NaPhosphate, containing 1.7 M ammonium sulphate, 0.05% CHAPS, and pH 7.4. Elute with a

30-mL inverse linear ammonium sulphate gradient from 1.7 *M* to 0 in 40 m*M* NaPhosphate, 1% CHAPS, pH 7.4.

4. Pool active fractions and dialyze into PBS/0.05% CHAPS.

3.5.10.2. Purification by Preparative SDS-PAGE

An alternative method for purification of CD59 from erythrocytes is preparative SDS-PAGE. Proteins can be highly purified from a relatively crude membrane preparation in a single step *(39)*. CD59 is remarkably resistant to denaturation and hence this method is suitable to purify functionally active CD59.

1. Dialyze butanol extract from 100 mL packed erythrocytes against PBS/0.1% NP40. Concentrate to 2.5 mL in an Amicon ultrafiltration cell using a 100-kDa cutoff membrane (CD59 has a M_r of 18–20 kDa on SDS-PAGE, but in NP40 resides in a high-molecular weight micelle).
2. Mix 1:1 with SDS-PAGE nonreducing sample buffer and load on a 15% SDS-PAGE tube gel (30 mL gel volume; Prep cell SDS-PAGE apparatus, Bio-Rad).
3. Run at 40 mA constant current and elute at 1 mL/min collecting 5 mL fractions. Dialyze fractions two times against PBS/0.05% CHAPS to remove SDS and assay for functional activity.
4. Pool and dialyze active fractions against PBS/0.05% CHAPS, concentrate in Amicon cell using a 10-kDa membrane and run on a Superose 12 column equilibrated in the same buffer. Pool active fractions and store at 4°C.

3.5.10.3. Functional Assay

This assay is based on the ability of CD59 to reincorporate into a cell through its GPI-anchor. Guinea pig erythrocyte (GPE) are the ideal target cells. Chicken E are an acceptable alternative but sheep E, pig E, or rabbit E are unsuitable, due to the high level of endogenous CD59 on these cells *(37,38)*. Fractions to be tested have to be in CHAPS because other detergents will lyse the erythrocytes. E can be incubated with up to 0.3% CHAPS without lysis.

1. In a 96-well plate, incubate 50 µL of a 2% suspension of GPE in PBS/0.05% CHAPS with 50 µL of column fraction, 30 min 37°C.
2. Wash cells twice in VBS to remove unincorporated proteins.
3. Incubate with 100 µL human serum 1/10–1/20 (serum dose chosen to cause near-100% lysis in absence of inhibitor) and CVF (1 µg/mL final conc.) 30 min 37°C.
4. Measure absorbance at 412 nm in supernatants. Identify fractions causing inhibition of E, pool and take to next step (*see* **Note 5**).

3.6. Purification of C Receptors

The first step in purification of a C receptor is to obtain a membrane extract from an appropriate cell type as described in **Subheading 3.5.7.** The appropriate cell source for a C receptor is determined by expression levels. There are no functional assays available for the majority of complement receptors other than binding of the relevant ligand. The purification methods described in this chapter are based on the affinity of the receptors for their ligands, ensuring the purification of the desired component. All C receptors have transmembrane anchors. Specific receptors have been described for C1q, fragments of C4, fragments of C3, and fragments of C5. Purification protocols for the majority of these have been described, most studied have been receptors for fragments of C3 (+/– the equivalent fragments in C4). Purification of CR1, CR2, CR3, and CR4 can all be performed by affinity chromatography using the appropriate C3 fragments as ligand. Differential elution of C receptors that have affinity for the same ligand can be achieved by removal of bivalent cations. For example, CR3 and CR4 elute from an iC3b column upon removal of bivalent cations, while CR1 and CR2 are eluted with 130 and 300 mM NaCl, respectively *(48)*. Purification of C3a receptor has not been described. The structure of the C3aR is very similar to that of C5aR and purification can probably be carried out using a similar protocol and specific assays (*see* below).

3.6.1. Purification of Receptors for C1q

At least three different receptors for C1q have been identified so far *(40–42)*. The two methods outlined here describe, respectively, the purification of the first putative C1qR identified, the so-called collectin receptor *(43)*, and the more recently described, but much better characterized phagocytic C1q receptor C1qRp *(41)*.

3.6.1.1. Isolation of the Collectin Receptor (cC1qR, M$_r$ 55–70 kDa) *(40,44,45)*

1. Extract Raji cells (4 × 10^{10} cells) or tonsil cells with 1% NP40 in 10 mM NaPhosphate, 0.15 mM CaCl$_2$, 0.5 mM MgCl$_2$, 100 µg/mL SBTI, 5 mM iodo-acetamide, 2.5 mM PMSF, 20 µM 1,10-phenanthroline, 5 µg/mL pepstatin, pH 7.4. Incubate 30 min on ice, spin 30 min 13000g, 4°C, and take supernatant.
2. Apply cell lysate to C1q-Sepharose column (3 mL; 3 mg C1q bound per mL Sepharose CL-4B; C1q purified as described in **Subheading 3.1.1.**) equilibrated in 10 mM NaPhosphate, pH 7.4, containing 0.1% emulphogene BC720 and the above enzyme inhibitors. Elute with 1 M NaCl and dialyze protein-containing fractions against 10 mM NaPhosphate, 0.1% emulphogene BC720, 7.4.
3. Apply to MonoQ column equilibrated in the same buffer and elute with 50 mL linear gradient from 0 to 1 M NaCl. Identify positive fractions and concentrate by ultrafiltration.
4. Fractionate on a Superose 12 gel filtration column equilibrated in 50 mM NaPhosphate, 2 mM EDTA, 150 mM NaCl, 0.1% emulphogene BC720, pH 7.4. Identify positive fractions, concentrate by ultrafiltration, and store at 4°C.

3.6.1.2. ISOLATION OF THE PHAGOCYTIC C1qR (C1qRp; M_r 126 kDa) *(41)*

1. C1q, purified as described in **Subheading 3.1.1.**, is subjected to limited tryptic digestion and the collagenous tails separated from globular heads by gel filtration on Superose 12.
2. C1q collagenous tails (2 mg) are immobilized on 3 mL sepharose 4B (using CNBr-activated sepharose).
3. Extract U937 cells (4×10^{10} cells) or monocytes cells with 1% NP40 in 10 mM NaPhosphate, 0.15 mM CaCl$_2$, 0.5 mM MgCl$_2$, 100 µg/mL SBTI, 5 mM iodoacetamide, 2.5 mM PMSF, 20 µM 1,10-phenanthroline, 5 µg/mL pepstatin, pH 7.4. Incubate 30 min on ice, spin 30 min 13,000g, 4°C, and take supernatant.
5. Apply cell lysate to C1q(tail)-Sepharose column equilibrated in 10 mM NaPhosphate, pH 7.4. 0.05% NP40 and the above enzyme inhibitors. Elute with 1 M NaCl in the above buffer and dialyze protein-containing fractions against PBS, 0.05% NP40 (*see* **Note 6**).

3.6.1.3. FUNCTIONAL ASSAY

Column fractions are best assayed by immunochemical methods using antibodies specific for the various receptors The collectin receptor (cC1qR) has been assayed by monitoring inhibition of C1q activity for lysis of ShEA *(44)*.

1. Incubate column fractions with 50 ng C1q, 60 min 37°C.
2. Add 10 µL C1q depleted serum (containing 20 mM CaCl$_2$; generated by passage over IgG column) and 50 µL 2% sensitized ShE in a total volume of 150 µL, incubate 30 min at 37°C and measure lysis.

3.6.2. Purification of C5aR (CD88)

C5aR: M_r: 42 kDa, seven transmembrane domains *(46,47)*.
The purification of the C5aR is carried out based on its affinity for C5a.

1. Mix one volume of packed neutrophils (6×10^9 cells) with 1 vol 50 mM Hepes, 2% digitonin, 0.1 mM PMSF, 10 µg/mL chymostatin , 10 µg/mL leupeptin, pH 7.2, incubate on ice for 1 min and remove insoluble material by centrifugation (200,000g, 7 min).
2. Coupling of C5a to commercial resins has proven to be very inefficient. The following procedure has been used for coupling of C5a to solid phase.
   a. Add 120 µL tetrahydrophthalic anhydride (3.65 mg/mL in dioxane) to 12 mL C5a (1 mg/mL in 0.1 M Hepes, pH 8.0) and incubate 5 h RT.
   b. Add 86 µL maleic anhydride (78.4 mg/mL in dioxane) and incubate overnight.
   c. Adjust pH to 5.7 and incubate overnight.
   d. Adjust pH to 8.0 and concentrate on an Amicon YM5 membrane to 4 mg/mL.
   e. Add to 2 mL Affigel 10 and incubate overnight at 4°C.

f. Wash gel with 100 mM Hepes pH 8.0, block with 1 M glycine, 4 h, wash with 100 mM Hepes, pH 3.2, and incubate 72 h, 37°C.
    Store in 0.1 M Hepes, pH 7.2.
3. Incubate neutrophil extract with 2 mL C5a-affigel, equilibrated in 100 mM Hepes, 0.1% digitonin, 5 mM MgCl$_2$, pH 7.2, and incubate 48 h with mixing.
4. Wash with buffer, containing 1 M NaCl and with 100 mM Hepes, 0.1% digitonin, pH 7.2. Elute with 50 mM formic acid, 200 mM KSCN, 0.1% digitonin, pH 4.0, and desalt immediately on 10 mL Bio-Rad P-10 column in 100 mM Hepes, 0.1% digitonin, 5 mM MgCl$_2$, pH 7.2. Eluted material consists of three components: C5aR and the α and β subunit of the associated G-protein.

3.6.3. Purification of Complement Receptor 1 (CR1/CD35)

CR1: M$_r$: nonred, 190 kDa, red, 240 kDa, 30 SCR, TM-anchor.
Erythrocytes have only a low level of expression of CR1, but because of the ease with which they can be obtained and because membrane extracts are relatively easy to prepare, erythrocytes are the preferred source of CR1 *(49)*.

1. Dialyze a 1% NP40 extract obtained from 20 U (5 L) of human erythrocytes against 0.1% emulphogene BC720 (Sigma), 1 mM benzamidine, pH 7.0, and 5 vol, 5 mM NaPhosphate, 0.5 mM EDTA, 5 mM benzamidine, pH 7.8 (DE buffer).
2. Apply to DEAE-Sephacel column (5 cm × 85 cm) equilibrated in DE buffer. Elute with 8 L of a linear gradient from 0 to 200 mM KCl. Identify positive fractions and dialyze against 10 mM KPhosphate, 0.5 mM EDTA, 5 mM benzamidine, 0.1% emulphogene, pH 7.0 (start buffer).
3. Apply to C3$_{MA}$-Sepharose 4B column (1.5 cm × 20 cm); 7.5 mg C3$_{MA}$ coupled to CNBr-Sepharose 4B C3$_{MA}$ prepared as described in **Subheading 3.1.5.** Elute with 300 mL linear gradient 0 to 350 mM NaCl in start buffer.
4. Pool positive fractions, dialyze into start buffer, reapply to C3$_{MA}$-Sepharose column and elute as above.
5. Pool positive fractions, concentrate and store at –20°C.

3.6.3.1. FUNCTIONAL ASSAY

Assay using cofactor activity with C3$_{MA}$ as substrate as described in **Subheading 3.5.2.**

3.6.4. Purification of Complement Receptor 2 (CR2/CD21)

CR2: M$_r$: nonred, 110–120 kDa; reduced 145 kDa, SCR 15 or 16; TM-anchor *(50)*.

1. Extract tonsil lymphocytes (2×10^{10}, approx 40 tonsil pairs) at 10^8 cells/mL with 3% (w/v) Brij96 in 150 mM NaCl, 2 mM PMSF, 15 min , 4°C, spin 10,000g 5 min, and add 0.1% Triton X-100 to supernatant. Dialyze against 10 mM NaPhosphate, 1% NP40, 2 mM PMSF, pH 7.4.

2. Apply to C3d,g-Sepharose column (1.4 cm × 7 cm; 2.4 mg C3d,g/mL Sepharose) and wash sequentially with 30 mL 10 mM NaPhosphate, 1% NP40, pH 7.4 containing 80 mM NaCl, 130 mM NaCl, and 190 mM NaCl, respectively. CR2 elutes with 130 mM NaCl. Pool and dialyze against 10 mM NaPhosphate, 1% NP40, pH 7.4

3. Apply to DEAE-Sephacel column (2 cm × 7 cm) equilibrated in the same buffer. Elute with 20 mL of 150 mM NaCl.

3.6.5. Purification of Complement Receptor CR3 (CD11b/CD18 Heterodimer)

CR3: 2 chain, TM: CD11b: 165 kDa; CD18: 95 kDa *(51)*.

CR3 can be purified from neutrophils or an appropriate cell line. All buffers contain 1 µg/mL each of the following protease inhibitors: antipain, benzamidine, chymostatin, leupeptin, aprotinin, and pepstatin. Also contain 2 mM PMSF, 5 mM pefablock.

1. Extract cells at 2–5 × 10^7/mL in 20 mM Hepes, 2.5% nOGP (Sigma), 50 mM NaCl, 1 mM CaCl$_2$, 1 mM MgCl$_2$, 1 mM MnCl$_2$, pH 7.4 for 10 min on ice. Spin 12,000g, 10 min.
2. Add supernatant to 100 µL C3bi-beads (*see* **Note 6**) in the same buffer.
3. Incubate the cell lysate with C3bi-beads for 3 min at RT and wash with lysis buffer. Elute with 2.5% nOGP, 20 mM Hepes, 10 mM EDTA, 50 mM NaCl pH 7.4 at RT.

3.6.6. Purification of Complement Receptor CR4 (CD11c/CD18 Heterodimer)

CR4: 2 chain, TM: CD11c: 150 kDa; CD18: 95 kDa *(48)*.

1. Extract neutrophils or spleen cells with 1% NP40 in the presence of protease inhibitors as above.
2. Pass extract over iC3b-Sepharose column (2 mL; 2.2 mg iC3b/mL Sepharose), equilibrated with 1% NP40 in 10 mM NaPhosphate, 1 mM CaCl$_2$, 1 mM MgCl$_2$, pH 7.4 and elute with 10 mM NaPhosphate, 2 mM EDTA, pH 7.4 (elutes CR3 and CR4); elute other iC3b binding proteins with 500 mM NaCl.
3. Further separation of CR3 and CR4 can be achieved using MAb specific for CD11b and CD11c, respectively.

4. Notes

1. C1 can be reconstituted by incubation of purified C1q, C1r, and C1s (molar ratio 1:2:2) in VBS containing 5 mM CaCl$_2$. If activated C1r and C1s are required, prior to **step 5** above the column is incubated for 30 min at 37°C in the absence of NPGB and then eluted in VBS containing 10 mM EDTA. To separate MBL from MASP-1/2, dialyze against 100 mM NaAc, 200 mM NaCl, 5 mM EDTA pH 5.0, and gel filter on Sephacryl S-300 (1 cm × 45 cm) equilibrated in the same buffer.

No methods have been described to separate MASP-1 from MASP-2. Purification can be followed on SDS-PAGE with CBB staining or western blotting. MASP activity can be measured using the same esterolytic assay used for C1r and C1s, but substituting N-CBZ-L-leu (Sigma) as substrate. Because methylamine also inactivates C4, alternative pathway (AP) activation is required for measuring functional activity. Zymosan is added to drive AP activation and enhance the sensitivity of the assay.

2. CVF is a useful tool in C research because it can form with human factor B a C3/C5 convertase (CVF.Bb) that is resistant to inactivation by factor I and H, giving it a very long half-life (7 h compared to 1.5 min for C3bBb). The CVF.Bb complex can, therefore, cause rapid and uncontrolled activation of C that depletes C activity both in vitro and in vivo. Although not a C component, protocols for purification of CVF are for this reason included here. Whereas CVF from the cobra *Naja naja kaouthia* forms a convertase capable of cleaving both C3 and C5, that from *Naja haje* can only form a C3 convertase. The raw material for purifying CVF is freeze-dried cobra venom; this must be handled with extreme care because of the presence of cardiotoxins and neurotoxins which can be fatal if introduced into the blood stream. Toxins in waste products generated during purification should be neutralized with 4 *M* NaOH. Some cobra venoms contain more than one molecule that has an effect on the complement system *(25)*, so care has to be taken to choose an appropriate screening method. Acquiring crude cobra venom has recently become almost impossible in many European countries because the CITES Treaty regulating trade in endangered species has severely restricted harvesting and supply of cobra venom.

3. The position of elution of human C9 from DEAE sephacel can be predicted by inspection of the fractions; C9 begins to elute just after the ceruloplasmin peak that gives a blue color to the fractions. It is only necessary to assay fractions from the beginning of the blue peak to some 20 mL after disappearance of the blue color. To purify C5b6 with good yield from serum, acute phase serum (from patients with recent episodes of infections, sports injuries, surgery, or childbirth) can be used because it contains elevated levels of C5 and C6 but not of C7 (not an acute phase reactant). However, acute phase serum may be difficult to obtain; a simple alternative is to use serum depleted of C7 by passage over an anti-C7 column.

4. For the isolation of the anaphylatoxins in their active state from serum it is essential to block the serum exopeptidase carboxypeptidase N completely. In order to achieve this, the inhibitor 6-aminohexanoic acid is used at 1 *M*, together with 1 m*M* 2-mercaptomethyl-5-guanidinopentanoic acid or 2-mercaptomethyl-3-guanidino-ethylthiopropranoic acid *(28)*. An alternative approach to generate C3a and C5a and to circumvent the problems with inactivation by carboxypeptidases is to use purified C3 or C5 and induce cleavage on CVF-Sepharose as described in **Subheading 3.1.5.** The fragments can then be separated by gel filtration. Recombinant C5a can be purchased from Sigma.

5. S-protein in column fractions is best assayed by dot blot or in ELISA using anti-S-protein antiserum. Numerous functional assays for S-protein have been

described, involving either inhibition of C reactive lysis (**Subheading 3.3.6.**) or induction of fibroblast adhesion and spreading. S protein is a "sticky" protein with a tendency to dimerise when impure. The purification should be carried out as quickly as possible and glutathione added to prevent dimerization. Once pure, S-protein is quite stable. Inhibition of lysis may not be sufficiently sensitive for identification of DAF-containing fractions; inhibition of C3b deposition on ShEA can be used instead, although this is technically demanding and requires a source of C8-depleted serum to avoid E lysis. Specific inhibition of C8 and C9 by CD59 can be measured by incorporation into GPE bearing C5b-7 or C5b-8 sites and developing the assay using purified C8/C9 or human serum diluted in PBS/5 mM EDTA as a source of C8 and C9.

6. Isolation of gC1qR with M_r 33 kDa can be achieved based on the affinity of this receptor for the globular heads of C1q *(42)*. An affinity column is made with globular heads of C1q, generated by trypsin digestion as described in **Subheading 3.6.1.** and a cell extract from Raji cells applied. C3b-beads are generated by incubating 100 μL Sepharose CL4B-200 (Sigma) with 1 mL of human serum for 1 h at 37°C. Wash five times in PBS. Sepharose activates C and allows C3b deposition on the matrix. Alternatively, couple C3bi or C3$_{MA}$ to CNBr-Sepharose-4B (2 mg/mL).

References

1. Hammer, C. H., Wirtz, G. H., Renfer, L., Gresham, H. D., and Tack, B. F. (1981) Large-scale isolation of functionally active components of the human-complement system. *J. Biol. Chem.* **256,** 3995–4006.
2. Harrison, R. A. and Lachmann, P. J. (1996) Complement and complement receptors, in *Weir's Handbook of Experimental Immunology*, 5th ed. (W. Herzenberg L.A, D. M., Herzenberg, L. A., and Blackwell, C., eds.), Blackwell Science, vol. 2, pp. 74.1–79.11.
3. Wing, M. G., Seilly, D. J., Bridgman, D. J., and Harrison, R. A. (1993) Rapid isolation and biochemical-characterization of rat C1 and C1q. *Mol. Immunol.* **30,** 433–440.
4. Stemmer, F. and Loos, M. (1984) Purification and characterization of human, guinea-pig and mouse C1q by fast protein liquid-chromatography (FPLC). *J. Immunol. Meth.* **74,** 9–16.
5. Peitsch, M. C., Kovacsovics, T. J., and Isliker, H. (1988) A rapid and efficient method for the purification of the complement subcomponents C1r and C1s in zymogen form using fast protein chromatography. *J. Immunol. Meth.* **108,** 265–269.
6. Neoh, S. H., Gordon, T. P., and Roberts Thomson, P. J. (1984) A simple one-step procedure for preparation of C1-deficient human-serum. *J. Immunol. Meth.* **69,** 277–280.
7. Tan, S. M., Chung, M. C. M., Kon, O. L., Thiel, S., Lee, S. H., and Lu, J. H. (1996) Improvements on the purification of mannan-binding lectin and demonstration of its Ca^{2+}-independent association with a C1s-like serine protease. *Biochem. J.* **319,** 329–332.

8. Hessing, M., Paardekooper, J., and Hack, C. E. (1993) Separation of different forms of the 4th component of human-complement by fast protein liquid-chromatography. *J. Immunol. Meth.* **157,** 39–48.

9. Kerr, M. A. and Gagnon, J. (1982) The purification and properties of the 2nd component of guinea-pig complement. *Biochem. J.* **205,** 59–67.

10. Ling, M., Piddlesden, S. J., and Morgan, B. P. (1995) A component of the medicinal herb Ephedra blocks activation in the classical and alternative pathways of complement. *Clin. Exper. Immunol.* **102,** 582–588.

11. Tack, B. F. and Prahl, J. W. (1976) Third component of human complement: purification from plasma and physicochemical characterization. *Biochem.* **15,** 4513–4521.

12. van den Berg, C. W., van Dijk, H., and Capel, P. J. A. (1989) Rapid isolation and characterization of native mouse complement component-C3 and component-C5. *J. Immunol. Meth.* **122,** 73–78.

13. Sanchez-Corral, P., Anton, L. C., Alcolea, J. M., Marques, G., Sanchez, A., and Vivanco, F. (1989) Separation of active and inactive forms of the third component of human complement, C3, by fast flow liquid chromatography (FPLC). *J. Immunol. Meth.* **122,** 105–113.

14. Jessen, T. E., Barkholt, V., and Welinder, K. G. (1983) A simple alternative pathway for hemolytic assay of human-complement component-C3 using methylamine-treated plasma. *J. Immunol. Meth.* **60,** 89–100.

15. Gitlin, J. D., Rosen, F. S., and Lachmann, P. J. (1975) The mechanism of action of the C3b inactivator (conglutinogen-activating factor) on its naturally occurring substrate, the major fragment of the third component of complement (C3b). *J. Exper. Med.* **141,** 1221–1226.

16. Davis, A. E. and Harrison, R. A. (1982) Structural characterization of factor-I mediated cleavage of the 3rd component of complement. *Biochem.* **21,** 5745–5749.

17. Lachmann, P. J., Pangburn, M. K., and Oldroyd, R. G. (1982) Breakdown of C3 after complement activation—identification of a new fragment, C3g, using monoclonal-antibodies. *J. Exper. Med.* **156,** 205–216.

18. Davis, A. E., Harrison, R. A., and Lachmann, P. J. (1984) Physiologic inactivation of fluid phase C3b—isolation and structural-analysis of C3c, C3d,g (α-2d), and C3g. *J. Immunol.* **132,** 1960–1966.

19. Williams, S. C. and Sim, R. B. (1993) Dye-ligand affinity purification of human-complement factor-B and beta-2 glycoprotein-I. *J. Immunol. Meth.* **157,** 25–30.

20. Johnson, D. M. A., Gagnon, J., and Reid, K. B. M. (1980) Factor D of the alternative pathway of human complement. Purification, alignment and N-terminal amino acid sequences of the major cyanogen bromide fragments, and localization of the serine residue at th active site. *Biochem. J.* **187,** 863–874.

21. Gotze, O. and Muller-Eberhard, H. J. (1974) The role of properdin in the alternative pathway of complement activation. *J. Exper. Med.* **139,** 44–57.

22. Farries, T. C., Finch, J. T., Lachmann, P. J., and Harrison, R. A. (1987) Resolution and analysis of 'native' and 'activated' properdin. *Biochem. J.* **243,** 507–517.

23. Pangburn, M. K. (1989) Analysis of the natural polymeric forms of human properdin and their functions in complement activation. *J. Immunol.* **142,** 202–207.

24. Vogel, C. W. and Muller-Eberhard, H. J. (1984) Cobra venom factor—improved method for purification and biochemical-characterization. *J. Immunol. Meth.* **73,** 203–220.
25. Vogt, W. (1982) Factors in cobra venoms affecting the complement system. *Toxicon* **20,** 299–303.
26. Podack, E. R., Kolb, W. P., and Muller-Eberhard, H. J. (1976) Purification of the sixth and seventh component of human complement without loss of hemolytic activity. *J. Immunol.* **116,** 263–269.
27. Jones, J., Laffafian, I., and Morgan, B. P. (1990) Purification of C8 and C9 from rat serum. *Compl. Inflam.* **7,** 42–51.
28. Hugli, T. E., Gerard, C., Kawahara, M., Scheetz, M. E., Barton, R., Briggs, S., Koppel, G., and Russell, S. (1981) Isolation of three separate anaphylatoxins from complement-activated human serum. *Mol. Cellular Biochem.* **41,** 59–66.
29. van den Berg, C. W., Aerts, P. C., and van Dijk, H. (1992) C1–inhibitor prevents PEG fractionation-induced, EDTA-resistant activation of mouse complement. *Mol. Immunol.* **29,** 363–369.
30. Burge, J., Nicholson-Weller, A.,and Austen, K. F. (1981) Isolation of C4-binding protein from guinea-pig plasma and demonstration of its function as a control protein of the classical complement pathway C3 convertase. *J. Immunol.* **126,** 232–235.
31. Crossley, L. G., and Porter, R. R. (1980) Purification of the human complement control protein C3b inactivator. *Biochem. J.* **191,** 173–182.
32. Sim, R. B. and Discipio, R. G. (1982) Purification and structural studies on the complement-system control protein beta-1h (Factor-H). *Biochem. J.* **205,** 285–293.
33. Dahlback, B. and Podack, E. R. (1985) Characterization of human S-protein, an inhibitor of the membrane attack complex of complement—demonstration of a free reactive thiol-group. *Biochem.* **24,** 2368–2374.
34. Jenne, D. E. and Tschopp, J. (1992) Clusterin: the intriguing guises of a widely expressed glycoprotein. *Trends Biochem. Sci.* **17,** 154–159.
35. Seya, T., Turner, J. R., and Atkinson, J. P. (1986) Purification and characterization of a membrane-protein (gp45-70) that is a cofactor for cleavage of C3b and C4b. *J. Exper. Med.* **163,** 837–855.
36. Nicholson-Weller, A., Burge, J., Fearon, D. T., Weller, P. F., and Austen, K. F. (1982) Isolation of a human erythrocyte membrane glycoprotein with decay-accelerating activity for C3 convertases of the complement system. *J. Immunol.* **129,** 184–189.
37. van den Berg, C. W., Harrison, R. A., and Morgan, B. P. (1993) The sheep analog of human CD59—purification and characterization of its complement inhibitory activity. *Immunol.* **78,** 349–357.
38. van den Berg, C. W. and Morgan, B. P. (1994) Complement-inhibiting activities of human CD59 and analogs from rat, sheep, and pig are not homologously restricted. *J. Immunol.* **152,** 4095–4101.
39. van den Berg, C. W., Harrison, R. A., and Morgan, B. P. (1995) A rapid method for the isolation of analogs of human CD59 by preparative SDS-Page—application to pig CD59. *J. Immunol. Meth.* **179,** 223–231.

40. Malhotra, R. and Sim, R. B. (1989) Chemical and hydrodynamic characterization of the human-leukocyte receptor for complement subcomponent C1q. *Biochem. J.* **262,** 625–631.

41. Guan, E., Burgess, W. H., Robinson, S. L., Goodman, E. B., McTigue, K. J., and Tenner, A. J. (1991) Phagocytic cell molecules that bind the collagen-like region of C1q—involvement in the C1q-mediated enhancement of phagocytosis. *J. Biol. Chem.* **266,** 20,345–20,355.

42. Peerschke, E. I. B., Reid, K. B. M., and Ghebrehiwet, B. (1994) Identification of a novel 33-kDa C1q-binding site on human blood—platelets. *J. Immunol.* **152,** 5896–5901.

43. Malhotra, R. (1993) Collectin receptor (C1q receptor): structure and function. *Behring Inst. Mitteilungen.* 254–261.

44. Ghebrehiwet, B., Silvestri, L., and McDevitt, C. (1984) Identification of the Raji cell membrane-derived C1q inhibitor as a receptor for human C1q—purification and immunochemical characterization. *J. Exper. Med.* **160,** 1375–1389.

45. Ghebrehiwet, B., Lim, B. L., Peerschke, E., I. B., Willis, A. C., and Reid, K. B. M. (1994) Isolation, cDNA cloning, and overexpression of a 33-Kd cell-surface glycoprotein that binds to the globular heads of C1q. *J. Exper. Med.* **179,** 1809–1821.

46. Rollins, T. E., Siciliano, S., and Springer, M. S. (1988) Solubilization of the functional C5a receptor from human polymorphonuclear leukocytes. *J. Biol. Chem.* **263,** 520–526.

47. Rollins, T. E., Siciliano, S., Kobayashi, S., Cianciarulo, D. N., Bonillaargudo, V., Collier, K., and Springer, M. S. (1991) Purification of the active C5a receptor from human polymorphonuclear leukocytes as a receptor-Gi complex. *Proc. Natl. Acad. of Sci. USA* **88,** 971–975.

48. Micklem, K. J. and Sim, R. B. (1985) Isolation of complement-fragment-iC3b-binding proteins by affinity- chromatography—the identification of p150,95 as an iC3b-binding protein. *Biochem. J.* **231,** 233–236.

49. Sim, R. B. (1985) Large-scale isolation of complement receptor type-1 (CR-1) from human-erythrocytes—proteolytic fragmentation studies. *Biochem. J.* **232,** 883–889.

50. Micklem, K. J., Sim, R. B., and Sim, E. (1984) Analysis of C3-receptor activity on human lymphocytes-B and isolation of the complement receptor type-2 (Cr-2). *Biochem. J.* **224,** 75–86.

51. van Strijp, J. A. G., Russell, D. G., Tuomanen, E., Brown, E. J., and Wright, S. D. (1993) Ligand specificity of purified complement receptor type-3 (CD11b Cd18, alpha(M)beta(2), Mac-1)—indirect effects of an Arg-Gly-Asp (Rgd) sequence. *J. Immunol.* **151,** 3324–3336.

3

Immunoaffinity Methods for Purification of Complement Components and Regulators

B. Paul Morgan

1. Introduction

Immunoaffinity protocols offer a rapid, efficient, and simple way of obtaining a protein of interest. Almost all of the complement components, receptors, and regulators have been successfully isolated by immunoaffinity protocols. The use of these methods is limited only by the availability of suitable antibodies in sufficient quantities *(1–6)*. Immunoaffinity methods are compatible with most detergents used in the solubilization of membrane proteins and are thus well suited to the purification of membrane regulators of complement and complement receptors *(3,7,8)*. Proteins isolated by immunoaffinity methods, particularly on monoclonal antibody solid phases, are usually of high purity and require either no downstream processing or simple "polishing" steps, such as gel filtration to remove aggregates. Elution of the bound protein from the antibody-solid phase is commonly achieved by subjecting the column to extremes of pH to disrupt antibody–antigen interaction. Some complement proteins are labile at these pH extremes and methods must be modified accordingly (**Table 1**). Here, I will describe the steps followed in establishing an immunoaffinity protocol for a complement protein. These differ little from those used for any other target protein. Example protocols will be provided for a serum complement protein and a membrane bound complement regulator.

1.1. Choosing an Appropriate Antibody for Immunoaffinity Purification

The first step in establishing any immunoaffinity protocol is to obtain a suitable antibody. Suitable in this context implies that the antibody will bind the target protein with a sufficiently high affinity to retain the protein on the solid

From: *Methods in Molecular Biology, vol. 150: Complement Methods and Protocols*
Edited by: B. P. Morgan © Humana Press Inc., Totowa, NJ

Table 1
Modifications of Basic Protocol for Specific Components

Component	Problem	Solution
C1	Dissociates in absence of divalent cations	Avoid chelating agents in serum/plasma, use VBS instead of PBS, avoid chaotropes, include protease inhibitor NPGB (1 mM).
C2/fB	Very prone to proteolysis/denaturation	Ensure all steps are performed at 4°C, include protease inhibitors in plasma and buffers.
C3/C4	Very abundant in plasma, column saturates	Reduce the volume of plasma used–20 mL plasma should yield >10 mg C3!
fD	Very low concentrations in plasma	Consider alternative sources—urine from patients with renal pathology?
Properdin	Tendency to form aggregates	Minimize time of procedure, keep at 4°C, include protease inhibitors.
C8	Noncovalently associated subunits ($\alpha\gamma/\beta$) dissociate in high salt	Avoid high salt wash (250 mM max), use solid phase which binds both subunits (mixed MAb).
C9	Labile at pH extremes	Elute in chaotropes (e.g., 2 M MgCl$_2$) or minimize time of exposure by neutralizing immediately.

phase through washing procedures that remove nonspecifically bound serum proteins. On the other hand, the affinity of the antibody should be low enough to permit the efficient elution of the bound protein under conditions that do not cause denaturation. The antibody itself should be stable throughout the procedure and be retained on the column. Leaching of antibody from the affinity column is a common problem that can be minimized by choosing an appropriate antibody. A good immunoaffinity column will last for years! The amount of antibody required will be dictated by how much of the complement protein is required to be purified. Small affinity columns with a few milligrams of antibody bound are often adequate for research purposes, but for isolating large amounts of a particular component, columns bearing hundreds of milligrams of antibody may be required. Whenever possible, monclonal antibodies should be used for immunoaffinity columns. Purified polyclonal IgG can be used, but there are numerous disadvantages when compared to monoclonal solid phase:

- only a small proportion of the IgG in a good polyclonal antiserum is against the target antigen (typically 0.5–2%);
- antibodies against contaminants are likely to be present, resulting in copurification of contaminants and necessitating more downstream processing;

- polyclonal IgG solid phases are usually strong activators of complement and serum applied to the column must contain ethylenediaminetetraacetic acid (EDTA) to inhibit activation; monoclonal IgG solid phases are rarely strong activators.

2. Materials

2.1. Buffers

1. Coupling buffer for generation of antibody sepharose solid phases is 0.1 M NaHCO$_3$ pH 8.3, 0.5 M NaCl.
2. Blocking buffer is 0.2 M glycine pH 8.0.
3. Low pH wash buffer is 0.1 M Na acetate pH 4.0 containing 0.5 M NaCl.
4. Phosphate-buffered saline (PBS) is supplied in tablet form by Oxoid Ltd.
5. Neutralizing buffer is 1 M Tris-HCl pH 7.0.
6. Lysis buffer is 5 mM Na Phosph. pH 7.4 containing 2 mM EDTA, 1 mM benzamidine, 1 mM phenyl methyl sulfonyl fluoride (PMSF) and 0.05% NaN$_3$.

2.2. Column Matrices

1. Cyanogen bromide (CNBr)-activated Sepharose 4B (Amersham Pharmacia).
2. Sepharose 4B, protein A Sepharose (Amersham Pharmacia).

2.3. Equipment

1. Pumps, fraction collectors, and absorbance monitors for standard protein purification system (available from many manufactureres, including Bio-Rad and Amersham Pharmacia).
2. Automated or semiautomated chromatography systems are also widely available and can simplify preparations.
3. Spectrophotometer capable of reading absorbance in ultraviolet (UV) and visible ranges.
4. A cold room or cold cabinet in which to perform chromatographic steps.
5. Centrifuges: ranging from benchtop microfuges to large, floor-standing refrigerated centrifuges capable of processing 5-10 L of fluid in one round.
6. Ultrafiltration system for concentrating/dialyzing samples between chromatography steps. We use an Amicon ultrafiltration cell (Amicon, Inc.); numerous other systems are available.

3. Methods

3.1. Immobilizing Antibody on CNBr-activated Sepharose 4B

Antibody-Sepharose solid phases can be easily made using ready-to-use CNBr-activated Sepharose 4B, which can be purchased from Amersham Pharmacia. *See* **Note 1**.

1. Weigh out the amount of freeze-dried CNBR-Sepharose required to give the desired column volume. Each gram of freeze-dried powder will yield 3.5 mL of swollen gel.

2. Rehydrate and wash in 1 m*M* HCl on a sintered glass funnel (use 200 mL wash volume per gram). Final wash (50 mL/g) is in coupling buffer. Use gel within 1 h of reswelling.

3. Dialyze antibody to be coupled into coupling buffer at a final antibody concentration of 1–10 mg/mL.

4. Retain small aliquot for assessment of uptake.

5. Mix antibody with swollen gel at 1–2 mg antibody per mL swollen gel, incubate 2 h at room temperature with constant mixing. The manufacturer's protocol advises much higher concentrations of protein; our experience is that immunoadsorbents work better with rather lower antibody densities.

6. Spin (100*g*, 5 min), remove supernatant, and retain for measurement of residual antibody. Antibody uptake can be assessed by measurement of the residual antibody concentration in the supernatant. A good approximation can be obtained by measuring the absorbance at 280 n*M* before (**step 3**) and after (**step 5**) incubation with gel. Uptake should be better than 90% of added antibody.

7. Resuspend in blocking buffer to block remaining active groups. Incubate 2 h at room temperature with constant mixing.

8. Wash gel, either by centrifugation as above or on a sintered glass funnel, in three cycles of low pH wash buffer (50 mL) and high pH (coupling buffer, pH 8.0, 50 mL). Finally, wash into PBS pH7.4 containing 0.1% NaN_3 and store in this buffer.

9. The immunoadsorbent should be stored at 4°C prior to and between uses. The immunoadsorbent is at this stage ready to use. It is, however, recommended that, immediately prior to use, the column is washed with the buffer to be used for protein elution to remove any loosely bound antibody and re-equilibrated in PBS.

3.2. Other Antibody Solid Phases

Numerous other Sepharose-based and non-Sepharose-based solid phases have been used to make immunoadsorbents for protein purification. Other Sepharose-based solid phases include AH-Sepharose, which has free amino groups, and binds free carboxyls in the protein, CH-Sepharose, which has free carboxyl groups and binds free amino groups and Epoxy-activated Sepharose which couples to epoxy groups (all from Amersham Pharmacia). All three incorporate spacer arms that will increase the mobility of the bound protein and have frequently been used for coupling of small molecules and peptides. They offer no clear advantage over CNBr-Sepharose for immobilizing antibody. Protein A and protein G are bacterial products that bind the Fc region of IgG from most mammalian species. All antigen-biding sites are thus available in the bound antibody. Sepharose-immobilized protein A or protein G (Amersham Pharmacia, many other manufacturers) can be used to rapidly and efficiently generate an immunoadsorbent matrix.

Magnetic solid phases have been used to provide a very rapid column-free method for immunaffinity purification *(10)*. Magnetic beads (Dynal or other sources) are directly or indirectly coated with specific antibody, incubated with the serum or cell extract, washed on the magnet, and specifically bound protein eluted. These methods are generally suitable only for small-scale preparations (a few micrograms).

Silica-based solid phases offer a practical alternative to Sepharose for large-scale purification. Antibody binds strongly to octadecyl silica, a standard matrix for high-pressure liquid chromatography (HPLC), and the immuno-adsorbent so formed can be used to isolate specifically bound proteins with high efficiency *(11)*.

3.2.1. Immobilizing Antibody on Protein-A Sepharose

1. The antibody to be immobilized is applied to the protein A-Sepharose column to saturate all available sites. The column is washed to remove unbound antibody.
2. Sample is applied and the column washed to remove unbound protein.
3. Bound protein is eluted using conditions chosen to avoid removal of antibody from protein A. This usually involves elution in high salt at neutral pH.
4. The immunoadsorbent can either be retained for future use or antibody eluted from the column by washing at low pH, freeing the immobilized protein A for other applications.

3.3. General Protocol for Immunoaffinity Purification of a Complement Component from Serum

This protocol is based upon published methods for purification of C9 and C8 *(12,4)*. All steps are carried out at 4°C and all buffers are prechilled.

1. Prepare column containing 20 mL antibody-sepharose solid phase as described above, wash in elution buffer (PBS containing 50 mM diethylamine, pH 11.5; *see* **Note 2**) and equilibrate in PBS. Also prepare a precolumn containing 20 mL underivatized Sepharose 4B, washed thoroughly into PBS (**Note 3**).
2. Centrifuge fresh serum or EDTA plasma (200–500 mL) 2000g, 15 min to remove insoluble material and apply to the precolumn (to bind proteins sticking nonspecifically to Sepharose) followed by the immunoaffinity column, placed in series, at a flow rate of 1–2 mL/min.
3. Wash columns in series with 50 mL PBS, disconnect precolumn, and wash immunoaffinity column with 100 mL PBS containing 0.5 M NaCl (**Note 3**).
4. Elute bound protein in two column volumes of elution buffer at a flow rate of 1 mL/min. Collect 1 mL fractions of the eluate into 0.1 mL neutralizing buffer to effect rapid neutralization of the pH. Wash column back into PBS containing 0.1% NaN$_3$ for storage.
5. Identify eluted protein in fractions by mixing 20-µL aliquots with 50-µL Coomassie protein assay reagent (Bio-Rad) in the wells of a microtiter plate. Pool protein peak and dialyze into PBS or other appropriate buffer. Concentrate in an Amicon ultrafiltration cell and store in aliquots at –20°C or below.

3.4. General Protocol for Immunoaffinity Purification of a
Complement Regulator or Receptor From Cell Membranes

This protocol is based upon published methods for purification from erythrocytes of sheep CD59 *(13)* and pig MCP *(14)*. The extraction procedure may need to be modified if the protein is to be purified from nucleated cells, although NP40 extraction as described below is usually successful. For glycosyl phosphatidylinositol (GPI) anchored complement regulators (DAF, CD59), butanol extraction of erythrocyte membranes is frequently used. In our experience, yields from Nonidet-P40 (NP40; Calbiochem) extracts are higher.

1. Obtain 500 mL packed erythrocytes, wash four times in 2 L PBS to remove all plasma, and buffy coat containing leukocytes.
2. Lyse erythrocytes by diluting in 10 L ice-cold lysis buffer and stir at 4°C for 1 h.
3. Pellet the erythrocyte ghosts by centrifugation ($25,000g$, 30 min, 4°C) and remove supernatant taking care not to disturb loose ghost pellet. Wash pellet by centrifugation, twice in lysis buffer and twice in PBS. Ghosts at this stage should be pale pink. If still red, repeat washing steps. Resuspend to total volume of 200 mL in PBS. Ghosts can be stored in PBS containing 0.1% NaN_3 for several days.
4. To the ghost suspension, add with constant mixing 50 mL of a 10% solution of NP40 in PBS to obtain a final detergent concentration of 2%.
5. Stir 30 min, 4°C, centrifuge as above to remove insoluble material.
6. Equilibrate precolumn and immunoaffinity column in cold PBS/0.2% NP40 and apply NP40 extract to the columns in tandem, essentially as described in Protocol **3.3.** above (**Note 4**).
7. Wash columns in series with 50 mL PBS/0.2% NP40, disconnect precolumn, and wash immunoaffinity column with 100 mL PBS/0.2% NP40 containing 0.5 M NaCl (**Note 5**).
8. Elute bound protein in two column volumes of elution buffer (50 mM diethylamine pH 11.5 in PBS/0.1% NP40), collect 1-mL fractions of the eluate into 0.1 mL neutralizing buffer to effect rapid neutralization of the pH. Wash column back into PBS containing 0.1% NaN_3 for storage.
9. Identify eluted protein in fractions by mixing 20-μL aliquots with 50 μL Coomassie protein assay reagent (Bio-Rad) in the wells of a microtiter plate (**Note 6**). Pool protein peak and dialyze into PBS/0.1% NP40 or other appropriate buffer. Concentrate in an Amicon ultrafiltration cell and store in aliquots at –20°C or below.

4. Notes

1. It is possible to CNBr activate Sepharose in the laboratory and simple protocols for this are available *(9)*. However, the difficulties inherent in handling CNBr make this an unattractive option unless it is planned to make very large quantities of antibody-solid phase. Cyanogen bromide reacts with hydroxyl groups on Sepharose and creates reactive groups that bind primary amine groups in proteins and peptides to form stable isourea linkages.

2. The elution conditions used are dictated by the stability of the protein and the ease of elution. Most involve extremes of pH (Tris/glycine pH 2.0 or diethylamine pH 11.5) or chaotropic agents ($2\,M$ $MgCl_2$, $4\,M$ NaSCN, $8\,M$ urea). For a new antibody/protein test several elution conditions for efficiency of purification and retention of function. Prior to elution, the column is washed with the highest salt concentration compatible with retention of specifically bound protein; washing with $1\,M$ NaCl will elute more contaminants and yield a cleaner product.

3. Sepharose 4B is supplied as a suspension in 20% ethanol. It is essential to wash out the ethanol before the precolumn is connected to the immunoaffinity column.

4. For CD59 and DAF it is our experience that yields are increased if the immunoaffinity column is run at room temperature. This may not be the case for other regulators and receptors. If the column is run at room temperature, it is essential to carry out the steps quickly. It should be possible to complete a purification procedure within a few hours.

5. Wash with the highest salt concentration compatible with retention of specifically bound protein; washing with $1\,M$ NaCl will elute more contaminants and yield a cleaner product. At this stage, detergent exchange can be carried out to switch to a detergent compatible with cell-based assays. For CD59 and DAF we routinely wash the column with 50 mL PBS/0.05% CHAPS (Calbiochem) and include CHAPS in place of NP40 in all subsequent buffers.

6. NP40 will itself yield a blue color when mixed with the Coomassie assay reagent, making detection of the protein peak impossible. CHAPS is compatible with the assay, a further advantage of detergent exchange.

References

1. Morgan, B. P., Daw, R. A., Siddle, K., et al. (1983) Immunoaffinity purification of human complement component C9 using monoclonal antibodies. *J. Immunolog. Meth.* **64**, 269–281.

2. Fowler, S. R. and Kolb, W. P. (1984) Utilization of immunoaffinity and high performance liquid chromatography for the isolation of human component complement C5: mediation of human monocyte chemotaxis by trypsin activated C5. *Federat. Proc.* **43**, 1460–1461.

3. Nemerow, G. R., Siaw, M. F. E., and Cooper, N. R. (1986) Purification of the Epstein-Barr virus/C3d complement receptor of human B lymphocytes: antigenic and functional properties of the purified protein. *J. Virology* **58**, 709–712.

4. Abraha, A., Morgan, B. P., and Luzio, J. P. (1988) The preparation and characterization of monoclonal antibodies to human complement component C8 and their use in purification of C8 and C8 subunits. *Biochem. J.* **251**, 285–292.

5. Misasi, R., Huemer, H. P., Schwaeble, W., Solder, E., Larcher, C., and Dierich, M. P. (1989) Human complement factor H: An additional gene product of 43 kDa isolated from human plasma shows cofactor activity for the cleavage of the third component of complement. *Europ. J. Immunol.* **19**, 1765–1768.

6. Seya, T., Hara, T., Iwata, K., Kuriyama, S. I., Hasegawa, T., Nagase, Y., et al. (1995) Purification and functional properties of soluble forms of membrane

cofactor protein (CD46) of complement: Identification of forms increased in cancer patients' sera. *Intl. Immunol.* **7,** 727–736.

7. Davies, A., Simmons, D. L., Hale, G., Harrison, R. A., Tighe, H., Lachmann, P. J., and Waldmann, H. (1989) CD59, an LY-6–like protein expressed in human lymphoid cells, regulates the action of the complement membrane attack complex on homologous cells. *J. Experiment. Med.* **170,** 637–54.
8. Purcell, D. F. J., Deacon, N. J., Andrew, S. M., and McKenzie, I. F. C. (1990) Human non-lineage antigen, CD46 (HuLy-m5): Purification and partial sequencing demonstrates structural homology with complement-regulating glycoproteins. *Immunogenetics* **31,** 21–28.
9. Axen, R., Porath, J., and Ernback, S. (1967) Chemical coupling of peptides and proteins to polysaccharides by means of cyanogen halides. *Nature* **214,** 1302–1304.
10. Karlsson, G. B. and Platt, F. M. (1991) Analysis and isolation of human transferrin receptor using the OKT-9 monoclonal antibody covalently crosslinked to magnetic beads. *Analyt. Biochem.* **199,** 219–222.
11. Chiong, M., Lavandero, S., Ramos, R., Aguillon, J. C., and Ferreira, A. (1991) Octadecyl silica: A solid phase for protein purification by immunoadsorption. *Analyt. Biochem.* **197,** 47–51.
12. Morgan, B. P., Daw, R. A., Siddle, K., Luzio, J. P., and Campbell, A. K. (1983) Immunoaffinity purification of human complement component C9 using monoclonal antibodies. *J. Immunolog. Meth.* **64,** 269–281.
13. van den Berg, C. W., Harrison, R. A., and Morgan, B. P. (1993) The sheep analogue of human CD59: purification and characterization of its complement inhibitory activity. *Immunology* **78,** 349–357.
14. van den Berg, C. W., Perez de la Lastra, J. M., Llanes, D., and Morgan, B. P. (1997) Purification and characterization of the pig analogue of human membrane cofactor protein (CD46/MCP). *J. Immunol.* **158,** 1703–1709.

4

Measurement of Complement Hemolytic Activity, Generation of Complement-Depleted Sera, and Production of Hemolytic Intermediates

B. Paul Morgan

1. Introduction

All of the basic functional assays of complement activity utilize erythrocytes as targets *(1)*. For classical pathway assays, the favored target is the antibody-sensitized sheep erythrocyte. For alternative pathway assays, unsensitized rabbit erythrocytes are routinely used. Numerous modifications on the basic assay methodology developed almost 50 years ago have found favor in different laboratories, which makes it difficult to compare results between laboratories. Some basic principles will be illustrated here and simple protocols for determination of hemolytic activities in the two major activation pathways (classical pathway CH50 and alternative pathway APH50) will be provided. It should be emphasized that measurements of total hemolytic complement provide only limited information. They are helpful as screening tests when complement deficiency is suspected (*see* Chapter 11) and can give a rather insensitive measure of complement activation. Despite this limited usefulness, hemolytic complement is often the only assay of complement activity available in the clinical laboratory. For accurate assessment of complement activity, serum samples for complement assay must be obtained fresh, promptly separated and either assayed immediately or stored frozen at −70°C until assay. Samples must not be subjected to freeze-thaw cycles.

Sera from which individual complement (C) components have been removed provide extremely useful reagents for assay of individual components and for generation of hemolytic intermediates. Methods can be broadly divided into "classical" methods utilizing chemical or thermal instabilities peculiar to particular components and immunoaffinity methods utilizing specific anti-C antisera or

From: *Methods in Molecular Biology, vol. 150: Complement Methods and Protocols*
Edited by: B. P. Morgan © Humana Press Inc., Totowa, NJ

monoclonal antibodies (mAb) coupled to sepharose. A few components can be removed from serum on nonimmunoglobulin solid phases (*see* **Note 1**).

2. Materials

1. Sheep erythrocytes (E) supplied in Alsever's solution by numerous suppliers, including TCS, Ltd. and Seralab (*see* Appendix for addresses).
2. Rabbit polyclonal antisheep E from Behring (Amboceptor).
3. Mouse monoclonal IgM antisheep E from (Sera-lab clone Spl HL).
4. Rabbit erythrocytes (RbE) supplied in Alsever's solution by numerous suppliers, including TCS, Ltd. and Seralab (see Appendix for addresses).
5. Phosphate-buffered saline (PBS).
6. Veronal-buffered saline (VBS).
7. Gelatin veronal buffer (GVB).
8. AP buffer is GVB containing 5 mM Mg^{2+} and 5 mM ethylene glycol-bis[β-aminoethyl ether]N,N'-tetraacetic acid (EGTA).
9. N-saline is 9 g NaCl dissolved in 1 L H$_2$O.
10. Barbitone buffer is made by mixing 0.1 M solutions of sodium barbitone and barbituric acid to obtain the target pH and adjusting volume to obtain required final molarity.
11. P-nitrophenyl p'-guanidinobenzoate (NPGB) from Sigma.

3. Methods

3.1. Preparation of Erythrocyte Targets for Hemolytic Assays

3.1.1. Antibody-Sensitized Sheep Erythrocytes (EA) for Classical Pathway Assay

1. Under aseptic conditions, remove 0.5 mL packed sheep E from stock of sheep blood, stored in Alsever's solution.
2. Wash E three times in PBS by centrifugation (2000g, 10 min, 4°C) and resuspend to 5% in PBS (10 mL), warm to 37°C in a water bath.
3. In a separate tube, warm a dilution of sensitizing antibody—rabbit polyclonal antisheep E or mouse IgM monoclonal antisheep E. The working dilution of antibody is titrated in preliminary experiments to find the maximum concentration, which does not cause agglutination of E.
4. Mix by pouring antibody solution into E suspension with constant agitation.
5. Incubate 30 min at 37°C with occasional agitation.
6. Wash EA twice in VBS and resuspend to original volume in VBS (5%, 10 mL) or GVB for immediate use.
   See **Note 2**.

3.1.2. Preparation of Rabbit Erythrocytes (RbE) for Alternative Pathway Assay

1. Under aseptic conditions, remove 0.5 mL packed RbE from stock rabbit blood stored in Alsever's solution.

2. Wash twice in AP buffer.
3. Resuspend in the same buffer to the required concentration for assay.
   See **Note 3**.

3.2. Measurement of Hemolysis in the Classical Pathway (CH50)

1. Make a 1:50 dilution of test and standard sera in GVB.
2. In the wells of a 96-well microtiter plate, make a further series of dilutions in GVB of each (1:50 diluted) serum sample 1:10; 1:5; 1:4; 1:3; 1:2; 1:1; 2:1; 3:1; 4:1; 5:1; 10:1; 10:0; 50 µL/well.
3. Add EA (in GVB at 10^9/mL), 50 µL/well.
4. Control wells in duplicate for each assay—100% contains 50 µL EA; cell blank contains 50 µL EA, 50 µL GVB.
5. Incubate 37°C 30 min with intermittent agitation.
6. To all serum wells and cell blanks add 150 µL ice-cold GVB, to 100% lysis wells add 200 µL H_2O. Spin plate, 1000g, 5 min.
7. Remove supernatants (200 µL) to new flat-bottomed 96-well plate, taking care not to disturb pelleted cells.
8. Read absorbance in supernatants at 540 nm, subtract cell blank absorbance from each value to obtain corrected absorbances (Ab^c).
9. Calculate fractional hemolysis in each well relative to the 100% lysis wells: Fractional hemolysis (y) = (Ab^c serum/Ab^c 100%).
10. Calculate the amount (in µL) of each serum giving 50% hemolysis (K) by plotting on log–log graph paper serum volume in µL added (x) vs [y/(1 – y)]; this will give a linear trace. At 50% hemolysis, $y/1 – y = 1$, hence the intercept on the x-axis from this point is *K*. *K* corresponds to 1 CH50 unit for that serum (**Fig. 1**).
11. The number of CH50 U/mL of the original serum, is obtained from: CH50 = 50 × (1000/K) U/mL. The first multiplier corrects for the 1:50 original dilution in the assay.
    See **Note 4**.

3.3. Measurement of Hamolysis in the Alternative Pathway (APH50)

1. Make a 1:25 dilution of test and standard sera in AP buffer.
2. In the wells of a 96-well microtitre plate, make a further series of dilutions in AP buffer of each (1:25 diluted) serum sample 1:10; 1:5; 1:4; 1:3; 1:2; 1:1; 2:1; 3:1; 4:1; 5:1; 10:1; 10:0; 50 µL/well.
3. Add RbE (in AP buffer at 10^8/mL), 50 µL/well.
4. Control wells in duplicate for each assay—100% contains 50 µL rabbit E; cell blank contains 50 µL RbE, 50 µL AP buffer.
5. Incubate 37°C 30 min with intermittent agitation.
6. To all serum wells and cell blanks add 150 µL ice-cold N-saline, to 100% lysis wells add 200 µL H_2O. Spin plate, 1000g, 5 min.
7. Remove supernatants (200 µL) to new flat-bottomed 96-well plate taking care not to disturb pelleted cells.

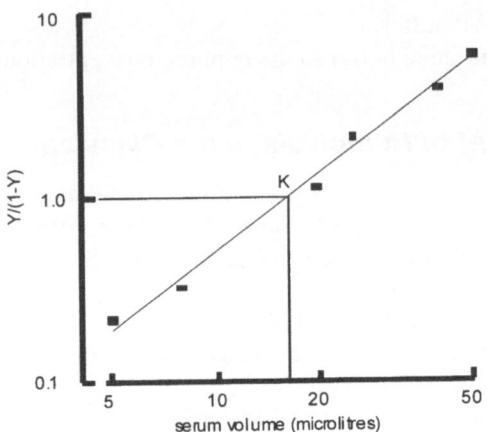

Fig. 1. Calculating CH50. The graph shows a log-log plot of $Y/(1 - Y)$ vs volume of diluted serum for a test sample. At the point of 50% hemolysis (K), $Y/(1 - Y) = 1$. The broken line from the Y-axis intercepts this point on the hemolysis curve and the intercept from this point to the X-axis permits the serum volume at 50% hemolysis to be read. In most laboratories these calulations are performed automatically by computer programs after inputting raw assay data.

8. Read absorbance in supernatants at 412 nm, subtract cell blank absorbance from each value to obtain corrected absorbances (Ab^c).
9. Calculate fractional hemolysis in each well relative to the 100% lysis wells: Fractional hemolysis (y) = (Ab^c serum/Ab^c 100%)
10. Calculate the amount (in μL) of each serum giving 50% hemolysis (K) by plotting on log–log graph paper serum volume in μL added (x) vs [$y/(1 - y)$]; this will give a linear trace. At 50% hemolysis, $y/1 - y = 1$, hence the intercept on the x-axis from this point is K. K corresponds to 1 APH50 unit for that serum.
11. The number of APH50 U/mL of the original serum, is obtained from: APH50 = 25 × (1000/K) U/mL. The first multiplier corrects for the 1:25 original dilution in the assay.

3.4. "Classical" Protocols for Generating Depleted Sera

3.4.1. Depletion of C1

C1 alone of the major C proteins is a euglobulin and can be removed from serum by dialysis against low-ionic-strength buffer.

1. Dialyze fresh serum (5 mL) overnight at 4°C against 2 L of 10 mM barbitone buffer pH 7.4 containing $CaCl_2$ (5 mM) and NPGB (0.1 mM).
2. Centrifuge serum at 5000g for 15 min at 4°C to remove the euglobulin precipitate.
3. Store C1-depleted serum in aliquots at –70°C.

3.4.2. Depletion of C4

The thioester group in C4 (and C3) is inactivated by treatment of serum with ammonia.

1. To 8.5 mL serum (guinea pig serum is most commonly used), add 1.5 mL NH_4OH diluted to 150 mM in H_2O.
2. Incubate for 45-min at 37°C; adjust pH to 7.4 with dilute HCl.
3. Store C4-depleted serum (R4) in aliquots at −70°C.

3.4.3. Depletion of C2

C2 and fB are the most heat labile of the C components *(3,4)*. Carefully controlled heating can be used to generate serum depleted of functional C2.

1. Place fresh serum (1 mL) in a glass tube preheated in a 56°C water bath and incubate at 56°C for precisely 6 min with constant shaking.
2. C2-depleted serum (R2) generated in this manner should be stored on ice and used immediately.

3.4.4. Depletion of C3

Incubation of serum with zymosan efficiently depletes C3 and partially depletes C5 and the terminal components. Zymosan depletion of C3 works better in guinea pig serum than in human serum where depletion is often incomplete.

1. Boil zymosan (100 mg; Sigma) in 10 mL VBS for 30 min. Wash twice in VBS and resuspend to 10 mg/mL in VBS.
2. Pellet 1 mL of zymosan suspension, remove supernatant, and suspend pellet in 10 mL serum.
3. Incubate 37°C 60 min; centrifuge (1000g 5 min) to pellet zymosan. Store supernatant (R3) in aliquots at −70°C.

3.4.5. Depletion of Factor B (fB)

1. Place fresh serum (1–2 mL) in a glass tube preheated to 50°C in a water bath.
2. Incubate for 20 min with continuous shaking.
3. fB-depleted serum generated in this manner should be stored on ice and used immediately.
   See **Note 5**.

3.4.6. Depletion of Factor D (fD)

fD can be selectively depleted from serum by virtue of its small size.

1. Apply fresh serum (1 mL) to a Sephadex G-75 gel filtration column (0.5 × 30 cm) equilibrated in VBS (fD alone of the C components is significantly retarded on this column).

2. Run the column in VBS and pool the void volume fractions, containing the bulk of the serum proteins minus fD. Store in aliquots at –70°C.

3.4.7. Depletion of Mannan-binding Lectin (MBL)

MBL can be depleted from serum utilizing its affinity for specific carbohydrates (see below).

3.4.8. Depletion of Terminal Pathway Components

Depletion of individual terminal pathway components (C5, C6, C7, C8, or C9) is best achieved by immunoaffinity methods using specific antisera or MAb (*see* below).

3.5. Depletion of C Components by Affinity Methods

3.5.1. General Protocol for Immunoaffinity Depletion of a C Component

1. Generate antibody solid phase using CNBr-activated Sepharose 4B (Pharmacia) exactly as described in **Chapter 3**. A column of 10 mL volume, with 20 mg monoclonal antibody bound or a column of 20 mL volume with 60 mg polyclonal antibody (purified IgG) bound should be adequate for most purposes.
2. Equilibrate immunoaffinity column in VBS at 4°C.
3. Allow buffer to run into solid phase until no buffer remains (but avoid drying out of column bed). Gently apply fresh serum (10 mL) using a pipet and taking great care not to disturb column bed. Allow serum to enter the solid phase and stop the flow through the column for 5 min.
4. Gently apply VBS (4°C) taking care not to disturb column bed and allow to flow slowly (0.5 mL/min) into solid phase.
5. Collect 0.5-mL fractions as serum is washed out of column. Discard obviously diluted fractions at beginning and end of elution and pool undiluted fractions.
6. Test completeness of depletion by incubating serial dilutions of the depleted serum (50 µL) with 50 µL of an appropriate indicator cell (sheep EA, 2% in VBS for classical pathway and terminal pathway components, rabbit E, 2% in APB for alternative pathway), 15 min at 37°C, centrifuging and measuring absorbance in supernatant at 412 n*M*. The depleted serum can be directly compared with the same serum before depletion; hemolytic activity should be reduced by >95%.
7. Regenerate immunoaffinity column by washing in two column volumes of 50-m*M* diethylamine pH 11.5 (elutes bound protein) followed by five column volumes of VBS; store in VBS containing 0.1% NaN$_3$.
8. If serum depletion is incomplete, reapply the depleted serum to the regenerated immunoaffinity column and repeat **steps 4–7**.
9. Test specificity of depletion by adding back physiological amounts of the missing component and demonstrating full restoration of lysis in the above assays.
10. Store depleted serum in aliquots at –70°C.
    See **Note 6**.

3.5.2. Depletion of C1q on Immobilized IgG (5,6)

1. Saturate a 5-mL column of Protein A Sepharose (Pharmacia) with IgG by applying purified human IgG (approx 100 mg applied slowly; Sigma) dissolved in VBS.
2. Wash column in VBS to remove unbound IgG.
3. Apply fresh human serum (20 mL) containing 10 mM ethylenediaminetetraacetic acid (EDTA), as described in **Subheading 3.5.1.**
4. Collect C1q-depleted serum, restore Ca^{2+} and Mg^{2+}, test, and store in aliquots at −70°C.
5. Regenerate column by washing with three column volumes of VBS containing 1 M NaCl and store in VBS containing 0.1% NaN_3. Eluate from column can be used as a source of C1q if NPGB is included in the elution buffer to inhibit activation.

3.5.3. Depletion of MBL on Sepharose 4B (7)

1. Equilibrate a 50-mL column of underivatized Sepharose 4B with VBS at 4°C.
2. Slowly apply 10 mL fresh human serum and elute in VBS.
3. Collect MBL-depleted serum, test, and store in aliquots at −70°C.
4. Elute bound MBL from column by washing with two column volumes of mannose (0.2 M) in VBS, wash and store column in VBS containing 0.1% NaN_3. *See* **Note 7**.

3.6. Generation of Hemolytic Intermediates

Hemolytic intermediates are target erythrocytes on which C activation has been permitted to proceed only as far as a predetermined point. The resulting intermediates can be used to assay for individual C components or to assess the activities of regulatory molecules. The depleted sera described above are valuable tools for the generation of hemolytic intermediates. Intermediates are often labile and have to be used immediately after generation.

3.6.1. Intermediates for Testing Classical Pathway Components and Regulators

1. EA are produced according to protocol **3.1.1.** above.
2. Incubate EA (2 mL, 2% in GVB) with purified human C1 (10–20 µg; Sigma or made as described in Chapter 2) at 30°C for 15 min with constant mixing. "Functionally pure" C1 will suffice.
3. Wash EAC1 by centrifugation and resuspend to 2% in GVB at 4°C. Store on ice for up to 6 h.
4. To 1 mL EAC1 on ice, add 1 mL of a 1:4 dilution of serum in GVB/10 mM EDTA prechilled on ice. Incubate with mixing on ice for 15 min.
5. Wash EAC14 three times by centrifugation in ice-cold GVB/EDTA and resuspend to 2%. Store on ice for up to 6 h. To form EAC4, these cells are incubated at 37°C for 30 min to allow removal of C1; EAC4 are stable for days at 4°C.
6. Incubate 1 mL EAC14 with 50 µg oxidized C2 30°C, 15 min (*see* **Note 8**).

7. Wash EAC14^{oxy}C2 by centrifugation and resuspend to 2% in GVB (*see* **Note 8**).
8. Warm 1 mL EAC14^{oxy}C2 to 30°C, add 100 µg C3 (Chapter 2), and incubate 30 min at 30°C.
9. Wash by centrifugation and resuspend at 2% in GVB/40 mM EDTA; incubate 1 h at 37°C to permit decay of C1 and C2.
10. Wash EAC43 three times in GVB and resuspend to 2%. Store at 4°C, stable for up to 1 wk.
    See **Note 8**.

3.6.2. Intermediates for Testing Terminal Pathway Components and Regulators

3.6.2.1. EAC1-7

1. Incubate EA (2% in GVB) with C8-deficient human serum or serum depleted of C8 on an antibody column as described above at a final dilution of 1:10 in GVB, 15 min, 37°C.
2. Wash EAC1-7 twice in GVB and resuspend to 2%; store at 4°C for up to 1 wk.
   See **Note 9**.

3.6.2.2. EAC1-8

1. Incubate EA (2% in GVB) with C9-deficient human serum or serum depleted of C9 on an antibody column as described above at a final dilution of 1:10 in GVB, 15 min, 37°C.
2. Wash EAC1-8 twice in GVB and resuspend to 2%; store at 4°C for up to 6 h.
   See **Note 9**.

3.7. Hemolytic Assays for Individual Components Utilizing Hemolytic Intermediates and/or Depleted Sera

3.7.1. C1

1. Generate EAC4 cells as described in **Subheading 3.6.1.**
2. Sera to be tested diluted in GVB^{2+} in range 1:10^4 to 1:10^6; duplicate 50-µL aliquots in wells of a 96-well plate.
3. Add 50 µL EAC4 (2% in GVB^{2+}), incubate 30°C, 15 min to form EAC14.
4. Add 50 µL C2 (5 µg/mL in GVB^{2+}) and incubate 15 min at 30°C to form EAC142.
5. Add 50 µL serum diluted 1:10 in VBS/10 mM EDTA as a source of late components and incubate 30 min at 37°C.
6. Control wells: cell blank 50 µL EAC4 in 150 µL GVB^{2+}; 100% lysis, 50 µL EAC4 in 150 µL H$_2$O.
7. Centrifuge plates, remove supernatants to fresh flat-bottomed 96-well plate, and read absorbance at 412 nM. Subtract cell blank absorbance from each value to obtain corrected absorbances (Ab^c).
8. Calculate fractional hemolysis in each well relative to the 100% lysis wells:
   Fractional hemolysis (y) = (Ab^c serum/Ab^c 100%).

9. Calculate "C1-dependent CH50" for each serum essentially as described in **Subheading 3.2.**

3.7.2. C4

1. Generate EAC1 cells as described in **Subheading 3.6.1.**
2. Sera to be tested diluted in GVB^{2+} in range 1:10^4 to 1:10^6; duplicate 50-μL aliquots in wells of a 96-well plate.
3. Add 50 μL EAC1 (10^8/mL in GVB^{2+}), incubate 30°C, 15 min to form EAC1.
4. Continue as in protocol **3.7.1.** from **step 4–step 8**, using EAC14 in controls.
5. Calculate "C4-dependent CH50" for each serum essentially as described in **Subheading 3.2.**

3.7.3. C2

1. Generate EAC14 cells as described in **Subheading 3.6.1.**
2. Sera to be tested diluted in GVB^{2+} in range 1:10^3 to 1:10^5; duplicate 50-μL aliquots in wells of a 96-well plate and prewarmed to 30°C.
3. Add 50 μL EAC14 (prewarmed, 10^8/mL in GVB^{2+}), incubate 30°C, for the time required for maximum formation of C42 sites (T_{max}; **Note 1**).
4. Continue as in protocol **3.7.1.** from **step 5–step 8**, using EAC14 in controls.
5. Calculate "C2-dependent CH50" for each serum essentially as described in **Subheading 3.2.**
   See **Note 10**.

3.7.4. C3

1. Generate EAC14^{oxy}2 as described in **Subheading 3.6.1.**
2. Sera to be tested diluted in GVB^{2+} in range 1:10^3 to 1:10^5; duplicate 50-μL aliquots in wells of a 96-well plate.
3. Add 50 μL EAC14^{oxy}2 (10^8/mL in GVB^{2+}) containing C5, C6, and C7 (0.1 μg each), incubate 37°C, 30 min to form EAC1-7.
4. Add 50 μL GVB^{2+} containing C8 and C9 (0.1 μg each), incubate 37°C 30 min.
5. Continue as in protocol **3.7.1.** from **step 6–step 8**, using EAC14^{oxy}2 in controls.
6. Calculate "C3-dependent CH50" for each serum essentially as described in **Subheading 3.2.**

3.7.5. Terminal Components

The components C5, C6, C7, C8, and C9 are best measured using the relevant depleted serum in excess. It is unusual to quantify terminal component hemolytic activity, a simple detection assay is often adequate.

1. To EA (10^8/mL in GVB^{2+}), add a dilution of the relevant depleted serum in GVB^{2+} which alone causes no lysis of EA (usually in range 1:5–1:20).
2. Sera to be tested diluted in GVB^{2+} in range 1:10^2–1:10^4; duplicate 50-μL aliquots in wells of a 96-well plate.

3. Add 50-μL aliquots of EA/depleted serum, mix, and incubate 37°C 30 min. Continue as in protocol **3.7.1.** from **step 6–step 8** using EA in controls.
4. Calculate "component-dependent CH50" for each serum essentially as described in **Subheading 3.2.**

4. Notes

1. The wide availability of human and animal sera genetically deficient in individual C components has to some extent supplanted the need for depleted reagents. However, deficient sera are expensive if purchased from commercial sources. Hence, for many laboratories, depleted sera remain useful reagents for research and clinical assays.
2. EA can be stored in Alsever's solution at 4°C for up to 2 wk without loss of activity. Wash into VBS or GVB and use as described above. Test each batch of EA by incubation (37°C, 30 min) with serial dilutions of serum to ensure that 100% lysis is achievable. The concentration of sheep EA can be calculated by lysing 0.1 mL of the stock EA in 2.9 mL H_2O and measuring the absorbance of the supernatant at 541 nm; absorbance of 0.37 equates to 1×10^9/mL original concentration.
3. The concentration of RbE can be calculated by lysing 0.1 mL of the stock RbE in 2.9 mL H_2O and measuring the absorbance of the supernatant at 412 nm; absorbance of 0.29 equates to 1×10^8/mL original concentration.
4. The manipulation required to calculate CH50s appears at first sight complicated. However, it is relatively simple to write programs to automatically calculate the CH50s direct from the absorbance readings. The normal range should be established locally and standard sera with known CH50 values used as calibrators in all assays. Several kits are available commercially to measure CH50. The Autokit CH50 (Wako, Osaka, Japan) based on the method of Yamamoto et al. *(2)*, uses antibody-sensitized liposomes entrapping an enzyme in place of EA to produce an assay system which can be adapted for automatic analyzers.
5. This procedure also inactivates C2 and partially inactivates C6 and C7.
6. All steps are carried out at 4°C to minimize activation of C on the column. Some immunoaffinity columns, particularly those on which polyclonal immunoglobulin has been immobilized, cause significant activation of C on the column, even at 4°C. To circumvent this, serum can by made 5 mM with EDTA prior to application to the column and Ca^{2+} and Mg^{2+} restored after the depletion step.
7. Most methods for purification of MBL utilize mannose or other sugars immobilized on Sepharose. Tan and coworkers *(7)* showed that underivatized Sepharose 4B binds MBL avidly, eliminating the need for immobilized sugars. The MBL-Sepharose interaction is Ca^{2+}-dependent; EDTA-containing serum thus cannot be used in this protocol.
8. Oxidation of C2 stabilizes its interaction with C4b in the C4b2a complex resulting in increased half-life of the extremely labile convertase. Oxidation is achieved by incubation of purified C2 with iodine as described *(3,8)*. EAC14^{oxy}C2 are labile and must be stored on ice and used within 1 h of production. EAC1 are

used for assay of C1 inhibitor; EAC4 are used for assay of C1; EAC14 for assay of C2; EAC14^{oxy}C2 for assay of C3. Cofactor and decay assays utilizing these intermediates are described in Chapter 6.

9. Similar intermediates can also be made on unsensitized E without deposition of early components by using CVF to activate C in the fluid-phase and deposit terminal complexes on the E. EAC5-7 can be made by incubating E with CVF and C8-depleted serum and EAC5-8 by incubation with CVF and C9-depleted serum. EAC1-8 will undergo slow, spontaneous lysis even at 4°C and thus cannot be stored long term. Efficiency of generation of EAC1-7 and EAC1-8 can be tested by incubation with EDTA-serum as a source of terminal components or, better, with the purified components. Complete hemolysis should be obtained within 15 min at 37°C.

10. C2, once activated, decays rapidly from the cell. Timings must thus be accurate to ensure that all wells are incubated for exactly the same time in **step 3**; this may require staggered addition of reagents. The T_{max} is calculated as described in Chapter 6.

References

1. Mayer, M. M. (1965) Complement and complement fixation., in *Experimental Immunochemistry*, 2nd ed. (Kabat, E. A. and Mayer, M. M., eds.), Charles C. Thomas, Springfield, pp. 133–240.
2. Yamamoto, S., Kubotsu, K., Kida, M., Kondo, K., Matsuura, S., Uchiyama, S., et al. (1995) Automated homogeneous liposome-based assay system for total complement activity. *Clin. Chem.* **41,** 586–590.
3. Kerr, M. A. and Porter, R. R. (1978) The purification and properties of the second component of human complement. *Biochem. J.* **171,** 99–107.
4. Ling, M., Piddlesden, S. J., and Morgan, B. P. (1995) A component of the medicinal herb ephedra blocks complement activation in the classical and alternative pathways. *Clin. Exp. Immunol.* **102,** 582–588.
5. Medicus, R. G. and Chapuis, R. M. (1980) The first component of complement. I. Purification and properties of native C1. *J. Immunol.* **125,** 390–395.
6. Schifferli, J. A. and Steiger, G. (1985) A simple two-step procedure for the preparation of the first component of human complement (C1) in its native form. *J. Immunolog. Meth.* **76,** 283–288.
7. Tan, S. M., Chung, M. C. M., Kon, O. L., Thiel, S., Lee, S. H., and Lu, J. (1996) Improvements on the purification of mannan-binding lectin and demonstration of its Ca^{2+}-independent association with a C1s-like serine protease. *Biochem. J.* **319,** 329–332.
8. Parkes, C., Gagnon, J., and Kerr, M. A. (1983) The reaction of iodine and thiol-blocking reagents with human complement components C2 and factor B. Purification and N-terminal amino acid sequence of a peptide from C2a containing a free thiol group. *Biochem. J.* **213,** 201–209.

5

Measurement of Complement Lysis of Nucleated Cells

O. Brad Spiller

1. Introduction

Lysis is the final end point of complement attack, indicating that the mechanisms acting to protect the cell have been overwhelmed. There is, of course, enormous biological significance in the deposition of complement fragments on cells in vivo because these serve as opsonins and opsonization is accompanied by the generation of anaphylatoxins and chemotaxins, often *without* lysing the cell. Measurement of the generation of anaphylatoxins, cell-bound complement fragments and hemolysis of erythrocytes will be described in other chapters. This chapter will focus on methods for measuring the complement-mediated lysis of nucleated cells in vitro. Nonnucleated cells, such as erythrocytes, lyse with relative ease when exposed to sensitizing antibody and complement. In the case of erythrocytes, this causes release of hemoglobin, which can easily be measured in the supernatant (Chapter 4). In comparison, measuring the lysis of living nucleated cells is more challenging. Nucleated cells, unlike erythrocytes, do not contain a natural chromophore. Nucleated cells have active budding processes to shed membrane attack complexes (MAC) from the membrane and they have ion-pumps that actively reclaim released ions, stabilizing the osmotic balance and preventing rupture of the plasma membrane.

Methods for measuring complement lysis of nucleated cells involve assessing the integrity of the plasma membrane following incubation with the complement source. There are two basic strategies for assessing membrane integrity: (1) measurement of release of a marker from within the cell, or (2) measurement of entry of a marker into the cell. The methods developed to

From: *Methods in Molecular Biology, vol. 150: Complement Methods and Protocols*
Edited by: B. P. Morgan © Humana Press Inc., Totowa, NJ

measure these aspects may utilize easily detected reporter molecules that are radioactive, colored, or fluorescent.

The two methods most commonly used in the past exemplify these basic strategies of assessment. (1) Release of radioactive chromium *(1,2)*: ^{51}Cr is taken up by and retained within live cells and the release of the radionuclide into the supernatant following complement attack is assessed by scintillation counting (i.e., marker released from the cell). (2) Trypan blue exclusion *(3)*: In the absence of membrane disruption, cells exclude high molecular weight vital dyes, such as Trypan blue. Cells appearing blue by phase microscopy following complement attack are counted as lysed (i.e., marker entering the cell). Both of these methods have limitations. Microscopic examination to count Trypan blue positive cells is tedious, time consuming, and inherently subjective. Measurement of ^{51}Cr release requires the use of radioisotope and is limited by the slow leaching of ^{51}Cr that occurs spontaneously from the cells after loading, resulting in a background release. Some other methods are more similar to the measurement of erythrocyte lysis by assessing release of haemoglobin. For example, release of the cytoplasmic enzyme lactate dehydrogenase (LDH) can be measured using a chromogenic substrate, but this requires quite complex enzyme kinetic measurements and cytosolic enzyme leakage can occur in the reversible phase of cell damage *(3,4)*.

Fluorescent dyes have inherent advantages over radioisotopes and vital dyes. They are safe to handle, avoid the necessary precautions associated with radioisotope use, offer high sample throughput and short assay processing time, and are suitable for measurement in a 96-well plate format. Despite these advantages, fluorescence-based methods have yet to find broad acceptance. The major problems have been high spontaneous release of the fluorescent dyes *(5)* or excessively slow release following attack, decreasing sensitivity leading to extended sample processing time *(6)*. Two fluorescent dyes are now beginning to gain broader acceptance for measurement of nucleated cell killing—calcein.AM and propidium iodide. This chapter will focus on the use of these agents.

2. Materials

1. Cell medium appropriate for the propagation of the cell (*see* **Note 1**).
2. 24-well plates (for adherent cells) or round-bottom 96-well plates (for nonadherent cells) [Purchased from Gibco BRL, Paisley, UK and ICN Chemical Ltd., Oxford, UK, respectively].
3. Calcein.AM [stock diluted in dimethyl sulfoxide (DMSO) at a concentration of 1 mg/mL] or propidium iodide (stock diluted in water at a concentration of 1 mg/mL). Both may be purchased from Molecular Probes, Inc. (Eugene, OR).
4. A 0.1% solution of Triton X-100 (Sigma, Poole, UK) in distilled water.

5. Veronal-buffered saline (VBS) to make up serum dilutions. This should be made fresh each day. Sold as complement fixation diluent tablets (1 tablet makes 100 mL) by Oxoid (Basingstoke, UK).
6. Fresh serum as a source of complement. Blood is collected by venepuncture into a glass tube and allowed to clot at room temperature (1 h) prior to separation of the serum by centrifugation (1200g, 20 min, 4°C). If not used fresh, serum is stored in aliquots at –70°C. Avoid freeze-thaw cycles.
7. Flat-bottom 96-well plates (ICN Chemicals, Ltd.) to read the samples in a fluorimeter or tubes suitable for flow cytometry.
8. Flow cytometry solution: PBS containing 1% bovine serum albumin and 15 mM EDTA.

3. Methods

3.1. Propidium Iodide Assays

3.1.1. General Principles

The propidium iodide (PI) exclusion method is in many ways similar to the trypan blue exclusion assay. PI, like its close molecular relative ethidium bromide (EtBr), belongs to a class of substances known as phenanthridinium intercalators *(7)*. This means they preferentially intercalate into DNA and RNA. PI and EtBr are both highly polar and water soluble and it is this polar nature that excludes these agents from entry across the cell membrane of viable cells. PI will enter the cell following disruption of the membrane and, once in the cell, bind tightly in the nucleus by intercalating into DNA. Binding to nuclear components concentrates the fluorescent dye and thus increases the apparent fluorescence by 20–100-fold. The stoichiometry of PI binding is 1 molecule of dye to 4–5 basepairs of DNA, hence lysed cells can become very brightly labeled. Extracellular PI fluoresces weakly, which means that analysis can be performed without the need for washing the cells. The excitation and emission wavelengths for PI make it suitable for visualization by fluorescence microscopy or flow cytometry.

Methods for assessing complement-mediated lysis by uptake of PI routinely use flow cytometry; live cells are dim and dead cells have a strong fluorescence in the red (FL2) range. It is necessary to use a standardized amount of PI for analysis to avoid having the fluorescence of dead cells exceed the scale maximum. PI is best suited to measuring the lysis of nonadherent cells following complement attack because disaggregation of adherent cells often leads to nonspecific PI uptake. Following attack, the cells are placed on ice and analysis must be carried out within an hour of complement attack. The cells cannot be fixed, and the flow cytometer will identify only cells that have not fragmented during the preparation for flow cytometry. It is possible to use the PI method with adherent cells if the cells are disaggregated prior to complement attack, but cells that readhere quickly to surfaces are not suitable for this method.

3.1.2. PI Exclusion Assay for Nonadherent Cells

1. If antibody sensitization is required, wash and resuspend the cells to a concentration of 10^6/mL in serum-free culture medium containing the optimum concentration of sensitizing and/or blocking antibodies. Incubate in the CO_2 incubator for 15 min. Pellet the cells (3 min at $1000g$), discard the supernatant and resuspend the cells to a concentration of 10^7/mL in VBS.
2. Place 0.1-mL aliquots of cells into wells of a round-bottomed 96-well plate. To triplicate sets of cells, add 0.1 mL of appropriate dilutions of serum in VBS. Incubate cells for 1 h at 37°C in a CO_2 incubator, then place on ice.
3. To each well, add 50 μL of 2 μg/mL PI, diluted from the 1 mg/mL stock in flow cytometry buffer, precooled on ice. Keep the plate on ice until analyzed by flow cytometry. Analysis should be carried out as soon as possible after addition of PI and always within 1 h of the complement challenge. Cells cannot be fixed, as fixation will permeablize all of the cells.
4. Measure cell fluorescence in the red range (FL2) on the flow cytometer. For measurement of PI uptake by flow cytometry, first set the parameters on the cytometer by running unattacked cells incubated with PI. Identify the cells by size (forward scatter) and granularity (side scatter) and adjust the baseline settings for FL2 to permit easy identification of cells showing an upward shift in fluorescence following complement attack. Samples are then run using gates appropriate to measure the percentage of the total cells with high fluorescence (lysed) for each sample.

3.2. Calcein Release Assay

3.2.1. General Principles

The calcein release method is the fluorescent equivalent of the ^{51}Cr-release assay. Calcein is an organic, polyanionic fluorochrome derived from fluorescein, but unlike other fluorescein-derivatives, its fluorescence is essentially independent of pH between 6.5 and 12 *(8)*. Historically, calcein was used to measure metal ion concentration because calcein fluorescence is quenched at physiological pH when bound to Fe^{3+}, Co^{2+}, Cu^{2+}, and Mn^{2+}, but not Ca^{2+} or Mg^{2+} *(9–11)*. The acetoxymethyl ester form of calcein (calcein.AM) is only weakly fluorescent and non-polar. The nonpolar nature of calcein.AM enables it to freely diffuse across membranes; once inside a living cell, the acetoxymethyl group is cleaved off by ubiquitous nonspecific cytoplasmic esterases *(12)*. The deesterified calcein is polar and is thus trapped in the cell until a breach of the plasma membrane occurs. Preloading of cells with calcein does not affect their ability to proliferate or interfere with other cellular functions *(13)*. Even erythrocytes or other nonnucleated cell types can be loaded with calcein. Methods using calcein.AM routinely use a fluorimeter to measure release of entrapped calcein. Calcein fluoresces in the green (FL1) range; therefore, calcein-loaded cells can also be analyzed by flow cytometry. Some assays for nonadherent cells combine calcein and ethidium homodimer (a close relative of PI) using two-color flow cytometry analysis or fluorescence microscopy: bright green cells are alive and red are dead *(14,15)*. A commercial kit based upon this dye combination, the LIVE/DEAD kit, is marketed by Molecular Probes, Inc.

3.2.2. Calcein Release Assay for Adherent Cells

1. Seed the cells evenly into the wells of a 24-well tissue-culture plate in order to obtain 70–90% confluence 24–48 h later (*see* **Notes 2–4**). The volume of cell medium should be no less than 0.5 mL per well. Sufficient wells should be seeded to enable each condition to be tested in triplicate so that statistical analysis can be performed. Control wells incubated in the absence of serum are included for measurement of background release of calcein.
2. Dilute the calcein.AM stock (1 mg/mL in DMSO) 1/500 in tissue-culture medium. Completely aspirate the medium from the cells and immediately add 0.25 mL of the diluted calcein.AM per well. Replace the plate in a CO_2 incubator at 37°C and leave to load for 1 h (*see* **Note 5**).
3. Aspirate the medium from the cells and add 0.2 mL/well of the appropriate sensitizing and/or blocking antibodies diluted in serum-free cell culture medium. Incubate at 37°C for 15 min.
4. Aspirate the medium from the cells. Wash once with 0.25 mL serum-free medium per well. Add 0.25 mL of serum diluted to the appropriate concentration in GVB^{2+} (VBS containing 0.1% gelatin). Incubate at 37°C for 1 h.
5. Carefully aspirate the medium from each well, taking care not to disturb the cell monolayer (complement-mediated release) and transfer into the wells of a prelabeled 96-well flat-bottomed plate (Plate 1). It is recommended that the pattern of well contents be arranged in the same manner between the 24-well plate and the 96-well plate. To the 24-well plate, add 0.25 mL of 0.1% Triton-X 100 per well and incubate for a further 15 min. At the end of the incubation, aspirate the liquid from each well of the 24-well plate into a second flat bottomed 96-well plate (detergent release; Plate 2).
6. Read plates 1 and 2 in a fluorimeter with the emission wavelength set to 530 nm and the excitation wavelength set to 480 nm. The photomultiplier tube (PMT) settings and lamp intensity should be set so that most of the samples have fluorescence that lies within the linear range of the instrument as stated by the manufacturer.
7. The % calcein released caused by complement attack is calculated for each well as:

$$\% \text{ release} = (\text{plate 1 value})/ (\text{plate 1 value} + \text{plate 2 value}) \times 100$$

Specific release is % release calculated above minus % calcein release in control wells not exposed to complement attack.

8. The total cell loading should be checked for each condition to ensure that there are not large differences in calcein uptake with different conditions. However, as the % release is calculated on a well-to-well basis, artefacts caused by differences in total loading or small differences in cell density between wells are likely to be small.

3.2.3. Calcein Release Assay for Cells in Suspension

1. Load the cells (10^7/mL) by incubating for 1 h at 37°C in a CO_2 incubator with calcein.AM (1 mg/mL stock in DMSO) diluted 1/500 in culture medium (*see* **Note 5**). Centrifuge for 3 min at 1000g, discard the supernatant, and resuspend the cells to the original volume in serum-free culture medium.
2. If antibody sensitization is required, add an appropriate volume of the sensitizing antibody to the cell suspension and incubate in the CO_2 incubator for 15 min.
3. Pellet the cells by centrifugation (3 min at 1000g), discard the supernatant, wash once in serum-free culture medium, and resuspend the cells in GVB^{2+} to the original volume (10^7/mL).
4. Place 0.1-mL aliquots (10^6 cells) into the wells of a 96-well plate, add 0.1 mL of serum dilution in VBS. Incubate the plate at 37°C for 1 h with occasional agitation.
5. Pellet the cells by centrifugation of the plate (3 min at 1000g) and carefully transfer all the supernatant to a prelabeled 96-well flat-bottomed plate (complement release; Plate 1), taking care not to disturb the cell pellet. Resuspend the cells in 0.2 mL 0.1% Triton-X 100 and incubate for a further 15 min. At the end of the incubation, transfer all of the well contents (0.1% Triton-X 100 and lysed cells) from the round-bottomed 96-well plate into a second flat-bottomed 96-well plate (residual calcein; Plate 2).
6. For measurement of calcein release by flow cytometry, cells from **step 4** above are diluted to 1 mL in ice-cold GVB^{2+} and kept on ice. The flow cytometer is calibrated by running unattacked calcein loaded cells. Calcein is visualized as green fluorescence (FL1) on the flow cytometer. The cells are identified by size (forward scatter) and granularity (side scatter) and the FL1 settings adjusted to permit easy identification of cells showing a significant reduction in fluorescence following complement attack. The percentage of the total cells retaining high fluorescence (unlysed) is measured for each sample. From this, percentage cell lysis is easily derived.

3.3. Optimizing Lysis for Specific Assays

Complement-mediated lysis may be measured for a variety of reasons. The target of an experiment may be to observe a reduction in lysis or an increase following a specific treatment and the protocol may need to be modified with this in mind. Systems designed to test decreased complement mediated lysis need to have a baseline lysis greater than 50%, but not so high that increased complement regulation would be masked by the excessive complement activation (i.e., less than 90%). Similarly, to test increased lysis, the baseline lysis must be measurable but low (under 25%) to ensure that the increase in complement activation or decrease in complement regulation causes a detectable increase in cell lysis. It is also important to include control cells that have not been exposed to complement to assess spontaneous lysis in the assay.

3.3.1. Optimizing Sensitization

Excessive complement activation is rarely a problem in lytic assays; reducing the serum dose or amount of sensitizing antibody (if used) will usually suffice. All serum dilutions should be made in an appropriate buffer containing 1% (w/w) bovine serum albumin or 0.1% gelatin to reduce background lysis. More often, the problem is that complement activation is insufficient to generate adequate amounts of complement-mediated lysis. Increased activation of the alternative pathway can be induced by adding zymosan, a powerful alternative pathway activator, which will cause deposition of alternative pathway convertases on adjacent cells; this works particularly well for adherent cells.

Activation of the classical pathway is best accomplished by preincubation of the test cells with a polyclonal serum raised against the same or a related cell type. Care must be taken in the generation of a sensitizing antiserum. If the aim of the test is to examine the effects of altered endogenous complement inhibitor expression, then the sensitizing antibody should not be reactive with these inhibitors. One strategy is to use a cell line that lacks glycolipid (GPI)-anchored proteins (some Raji-derived cell lines) or lack CD55 and CD59. It is important to test the binding of the sensitizing antibody to the test cells under all conditions, as incubation of cells with cytokines or other cell activators often results in dramatic alterations in sensitizing antibody binding.

3.3.2. Decreasing Complement Regulation

In order to assess the role of a particular complement regulator, it may be necessary to first block function of other complement regulators. High expression of CD59 often prevents detection of even substantial changes in complement activation or the effects of large alterations in expression of C3 regulators. CD59 function can be blocked by incubating the cells with a saturating concentration of a blocking monoclonal antibody prior to challenge with complement. Of the monoclonal anti-CD59 antibodies available, BRIC229 (International Blood Group Reference Laboratory [IBGRL], Elstree, UK) and MEM-43 (Harlan Sera-Lab, Sussex, UK) are known to block human CD59 function and are not complement activating. If high levels of DAF expression (such as those observed on HeLa cells) are suspected to be masking significant alterations in terminal complement pathway regulation, then a combination of BRIC110 and BRIC216 (IBGRL) can be used to block the function of DAF.

4. Notes

1. The presence of fetal calf serum (FCS) does not interfere with the loading of calcein.AM as long as the serum has been heat inactivated beforehand. If the cells to be tested are freshly isolated, and have not been propagated in culture medium, then RPMI medium containing 10% FCS is usually a good choice.
2. If the experimental design involves virus infection of the cells or incubation with cytokines or other cell activators, take into consideration the length of time required prior to complement challenge and the effects of the conditions on cell propagation.
3. The recovery period is important only if the method of subculture involves the use of enzymes to subculture the cells. If nonenzymatic methods are used for subculturing (i.e., EDTA/PBS) then it is only necessary to wait until the cells are firmly attached to the plate prior to calcein loading.
4. Experiments that are designed to test the protective nature of transfected cDNA that are under selective maintenance to make stable transfectants should have appropriate controls. Some selection reagents, such as Hygromycin B, have been shown to enhance the complement activating ability of cells even if removed several hours before attack. An appropriate control for this would be comparison with the same cell line tranfected in the same manner with an irrelevant cDNA.
5. Pretreatments that require the absence of FCS, will not affect the loading of calcein. However, ensure that the control cells are treated in the same manner as the pretreated cells.
6. Calcein is a fluorescein derivative and therefore is visualized in the FITC range (FL1). Because live cells will fluoresce more intensely than damaged cells, set the baseline fluorescence sufficiently high to allow a reduction in fluorescence to be easily observed.

References

1. Brunner, K. T., Mauel, J., Cerottini, J-C., and Chapuis, B. (1968) Quantitative assay of the lytic action of immune lymphoid cells on ^{51}Cr-labelled allogeneic target cells in vitro: inhibition by isoantibody and by drugs. *Immunology* **14,** 181–196.
2. Neville, M. E. (1987) ^{51}Cr-uptake assay. A sensitive and reliable method to quantitate cell viability and cell death. *J. Immunol. Meth.* **99,** 77–182.
3. Berry, M. N., Halls, H. J., and Grivel, M. B. (1992) Techniques for pharmacological and toxicological studies with isolated hepatocyte suspensions. *Life Sci.* **51,** 1–16.
4. Piper, H. M., Hutter, J. F., and Spieckermann, P. G. (1984) Relation between enzyme release and metabolic changes in reversible anoxic injury of myocardial cells. *Life Sci.* **35,** 127–134.
5. Kolber, M. A., Quinones, R. R., Gress, R. E., and Henkart, P. A. 1988 Measurement of cytotoxicity by target cell release and retention of the fluorescent dye biscarboxyethyl-carboxyfluorescein (BCECF). *J. Immunol. Meth.* **108,** 255–264

6. Brenan, M. and Parish, C. R. (1988) Automated fluorometric assay for T-cell cytotoxicity. *J. Immunol. Meth.* **112,** 121–131.
7. Waring, M. J. (1965) Complex formation between ethidium bromide and nucleic acids. *J. Mol. Biol.* **13,** 269–282.
8. Chiu, V. C. and Haynes, D. H. (1977) High and low affinity Ca^{2+} binding to the sarcoplasmic reticulum: use of a high-affinity fluorescent calcium indicator. *Biophys. J.* **18,** 3–22.
9. Sawahara, H., Goto, S., and Kinoshita, N. (1991) Double fluorescent labeling method used for a study on liposomes. *Chem. Pharm. Bull. (Tokyo)* **39,** 227–229.
10. Cabantchik, Z. I., Glickstein, H., Milgram, P., and Breuer, W. A. (1996) Fluorescence assay for assessing chelation of intracellular iron in a membrane model system and in mammalian cells. *Anal. Biochem.* **233,** 221–722.
11. Breuer, W., Epsztejn, S., and Cabantchik, Z. I. (1995) Iron acquired from transferrin by K562 cells is delivered into a cytoplasmic pool of chelatable iron(II). *J. Biol. Chem.* **270,** 24,209–24,215.
12. Dive, C., Cox, H., Watson, J. V., and Workman, P. (1988) Polar fluorescein derivatives as improved substrate probes for flow cytoenzymological assay of cellular esterases. *Mol. Cell. Probes.* **2,** 131–145.
13. Weston, S. A. and Parish, C. R. (1990) New fluorescent dyes for lymphocyte migration studies. Analysis by flow cytometry and fluorescence microscopy. *J. Immunol. Meth.* **133,** 87–97.
14. Poole, C. A., Brookes, N. H., and Clover, G. M. (1993) Keratocyte networks visualised in the living cornea using vital dyes. *J. Cell Sci.* **106,** 685–691.
15. Papadopoulos, N. G., Dedoussis, G. V., Spanakos, G., Gritzapis, A. D., Baxevanis, C. N., and Papamichail, M. (1994) An improved fluorescence assay for the determination of lymphocyte-mediated cytotoxicity using flow cytometry. *J. Immunol. Meth..* **177,** 101–111.

6

Functional Assays for Complement Regulators

Claire L. Harris

1. Introduction

As described in Chapter 1, the complement system comprises a battery of at least 20 components capable of immediate response to a foreign organism or cell, resulting in lysis or opsonization and phagocytosis. Uncontroled complement activation would result in damage to host tissues, or in extreme cases, consumption and effective depletion of complement components. Intrinsic to the complement cascade are multiple strategies for control, such as the inherent lability of the activation enzymes, the C3 and C5 convertases, and short-lived active/binding sites, such as the binding site on C5b67 for membranes. In addition to this inherent instability, complement activation is tightly controlled at multiple stages during the pathway by regulatory proteins, present both on cell membranes and in plasma *(1–4)*. The regulatory activities possessed by these proteins fall into four categories: (1) control of the active C1 complex; (2) "decay acceleration," characterized by the ability of the regulator to "break-up" the components of the multimolecular convertase enzymes; (3) "cofactor activity," enabling factor I to cleave and inactivate C3b or C4b; and finally, (4) inhibition of membrane attack complex (MAC) assembly. Some regulatory proteins have just one of these activities, others possess several.

1.1. C1 Inhibition

C1 activation is under the control of just one inhibitor, C1 inhibitor (C1inh). The function and assay of this regulator is described in depth elsewhere (Chapter 12). C1inh is a member of the serine protease inhibitor (SERPIN) family and functions by forming a complex with the active C1 subcomponents, C1r and C1s. Complex formation results in decay and inactivation of the C1 complex.

From: *Methods in Molecular Biology, vol. 150: Complement Methods and Protocols*
Edited by: B. P. Morgan © Humana Press Inc., Totowa, NJ

1.2. Decay Acceleration

This activity is possessed by several regulators, namely factor H (fH), C4b binding protein (C4bp), decay-accelerating factor (DAF), and complement receptor 1 (CR1). fH and C4bp are fluid phase regulators, whereas DAF and CR1 are located at the plasma membrane. The C3 convertase formed through activation of the alternative pathway (AP) comprises a complex of C3b and a cleavage product of factor B (fB), Bb. The classical pathway (CP) convertase consists of a complex between C4b and a cleavage product of C2, C2a; both convertases are dependant on Mg^{2+} for formation and are labile, with half-lives of a few minutes. The AP convertase can, however, be stabilized by interaction with another component of the pathway, properdin (*P*). Bb or C2a that has dissociated from the complex is inactive and cannot rebind. Decay-accelerators promote the dissociation of the convertase components, thereby decreasing the half-life of the complex even further. fH regulates the AP, it is particularly important for discrimination between activating and nonactivating surfaces, and for control of the AP feedback cycle. C4bp has an analogous role in the CP. DAF and CR1 regulate both pathways, although DAF is reported to regulate the CP more efficiently than the AP.

1.3. Cofactor Activity

The regulators with this activity are fH, C4bp, CR1 and a membrane-associated regulator, membrane cofactor protein (MCP). These four proteins act as essential cofactors for a plasma serine protease, termed factor I (fI). Association of a cofactor with C3b or C4b renders these large subunits of the convertase susceptible to cleavage and inactivation by fI. Cleavage of C3b at two sites in the α'-chain results in formation of iC3b, and release of a small 3-kDa fragment, C3f. One further cleavage in the α'-chain results in formation of *C3c* (145 ka) and C3dg (41 kDa), this latter fragment contains the thioester site. Cleavage of C4b at two sites in the α'-chain, one either side of the thioester, results in formation of the inactive fragments, C4c (147 kDa) and C4d (45 kDa). Cleavage patterns formed through inactivation of either protein are easily recognizable by SDS-PAGE. As is the case with decay activity, fH regulates only the AP, C4bp regulates only the CP, and CR1 can regulate both pathways. MCP is a cofactor for cleavage of both C3b and C4b, although it is reported to preferentially regulate the AP.

1.4. Inhibition of MAC Assembly

The terminal pathway is initiated by formation of C5b. This complement component subsequently binds C6 and C7, forming the C5b67 complex. Various fluid phase regulators work at this stage in the pathway by binding C5b-7 and preventing its association with membranes; these inhibitors include S-protein, clusterin, serum lipoproteins, and the complement component C8. If C5b-7 succeeds in associating with a membrane, it can bind C8 and multiple C9 resulting in formation of the MAC. Two further regulators have been described that act here, homologous restriction factor (HRF) and CD59. HRF is poorly characterized and its mechanism of action is unclear. CD59 on the other hand, is well characterized and acts by binding C8 in forming MAC and preventing incorporation of C9 into the pore.

2. Materials

2.1. Decay Assays

2.1.1. Buffers and Equipment

1. Shaking water bath at 30°C.
2. Shaking water bath at 37°C.
3. Centrifuge.
4. Spectrophotometer (414 nm).
5. Veronal-buffered saline (VBS) containing Mg^{2+} and Ca^{2+} (VBS^{2+}); supplied in tablet form as complement fixation diluent by Oxoid, Ltd.
6. Gelatin veronal buffer (GVB^{2+}): VBS^{2+} containing 0.1% (w/w) gelatin.
7. Dextran gelatin veronal buffer ($DGVB^{2+}$): 1 vol GVB^{2+} plus 1 vol 5% D-glucose in H_2O.
8. GVB-EDTA: VBS containing 0.1% gelatin and 40 mM ethylenediamine tetra-acetic acid (EDTA).
9. Alternative pathway buffer (APB): VBS containing 7 mM $MgCl_2$ and 10 mM ethylene glycol-*bis* (β-aminoethyl ether) N,N,N',N'-tetraacetic acid (EGTA).
10. Phosphate-buffered saline (PBS) supplied in tablet form by Oxoid, Ltd.
11. Saline solution: 0.9% (w/v) NaCl in H_2O.
12. Assay for C3a (ELISA from Quidel, Inc; RIA from Amersham Pharmacia).

2.1.2. AP Hemolytic Assay

1. Sheep erythrocyte intermediates bearing C4b and C3b ($E_{sh}AC4b3b$): (Chapter 4).
2. Purified properdin (*P*): (Chapter 2).
3. Purified factor D (fD): (Chapter 2).
4. Purified factor B (fB): (Chapter 2).
5. Rat serum diluted 1:20 in GVB-EDTA (C^{rat}-EDTA).

2.1.3. Classical Pathway (CP) Hemolytic Assay

1. Sheep erythrocyte intermediates bearing C1 and C4b ($E_{sh}AC14b$): (Chapter 4).
2. Purified C2: (Chapter 2).
3. Rat serum diluted 1:20 in GVB-EDTA (C^{rat}-EDTA).

2.1.4. Inhibition of Total C Activity

1. Antibody-coated sheep erythrocytes ($E_{sh}A$): (Chapter 4).
2. Rabbit E (RbE).
3. Serum.

2.1.5. C3a *Generation*

1. Purified components (CP): C4, C2, C3, C1s (Chapter 2).
2. Purified components (AP): C3, fB, fD, C3i (Chapter 2).

2.2. Cofactor Assays

2.2.1. Buffers and Equipment

1. Borate-buffered saline (BBS), pH 8.3: 0.1 M H_3BO_3 (boric acid), 0.025 M $Na_2B_4O_7 \cdot 10H_2O$ (borax), 0.075 M NaCl.
2. PBS.
3. Cell lysis buffer, PBS containing 1% NP40, 10 mM EDTA, 1 mM phenyl methyl sulfonyl fluoride (PMSF), 1 µg/mL pepstatin A, 1 µg/mL leupeptin.
4. Sephadex G25 column.
5. Reagents for SDS-PAGE (*see* Chapter 13).
6. Reagents for Western blotting, if required (see Chapter 13).
7. Water bath at 37°C.
8. γ-counter or phosphoimager.

2.2.2. α-Chain Cleavage

1. Purified C3 or C4: (Chapter 2).
2. Purified factor I (fI): (Chapter 2).
3. Methylamine.
4. Iodogen.
5. Chloroform.
6. ^{125}I.
7. 1 M KI.
8. NP40, if required.

2.3. MAC Inhibition Assays

2.3.1. Buffers and Equipment

1. APB (see above).
2. Saline solution: 0.9% (w/v) NaCl in H_2O.
3. Water bath at 37°C.
4. Centrifuge.
5. Spectrophotometer (414 nm).

2.3.2. Reactive Lysis Assay

1. Guinea pig erythrocytes (GPE).
2. C5b6 (Chapter 2).
3. Purified components: C7, C8, and C9 (Chapter 2).

2.3.3. Cobra venom factor (CVF)-Reactive Lysis Assay

1. GPE.
2. CVF from *Naja naja kaouthia* (Chapter 2).
3. Normal serum, or serum depleted of C8 or C9 (Chapter 4).
4. C8, C9, or serum diluted in PBS, 10 mM EDTA (C-EDTA) as required.

3. Methods
3.1. Decay-Accelerating Activity

There are a variety of assays used to assess decay-accelerating activity of regulatory proteins. Hemolytic assays can be devised such that they are specific for either activation pathway by using components, such as fB or C2, that constitute only the AP convertase or CP convertase *(5–8)*. Use of purified components enables sequential assembly of complexes on the cell surface and dissection of the point during the cascade at which a regulator functions. The fluid-phase "byproduct" of an active C3 convertase, C3a, generated using either a cell-based C activation system or purified components in a fluid phase assay, is easily quantitated by radio-immunoassay or ELISA *(9,10)*. These assays specifically monitor the decay of the convertase. Alternatively, less-specific assays can be used to detect "protection" of cells from C attack, in these cases there is no distinction between decay-acceleration and cofactor activity. Sensitivity of antibody-coated sheep E to whole C ± test sample can give an indication as to whether a sample contains a regulatory activity. In the case of the GPI-anchored protein, DAF, there is an added bonus in that incubation of this protein with E results in its incorporation into the cell membrane in a functionally active form. Once incorporated, cells can be washed and assessed for increased protection from C attack. Regulation of the C activation pathways, whether by decay-acceleration or cofactor activity, affects the amount of C3b and C3a generated during activation. The presence or absence of an inhibitor, such as DAF, can profoundly influence the levels of complement deposition and subsequent lysis. Measurement of cell surface C3 deposition is discussed in depth in Chapter 10 and measurement of nucleated cell lysis is described in Chapter 5. These two latter techniques are particularly useful when working with recombinant proteins expressed on the surface of transfected cells. A comparison of C3 fragment deposition and cell lysis between control cells and

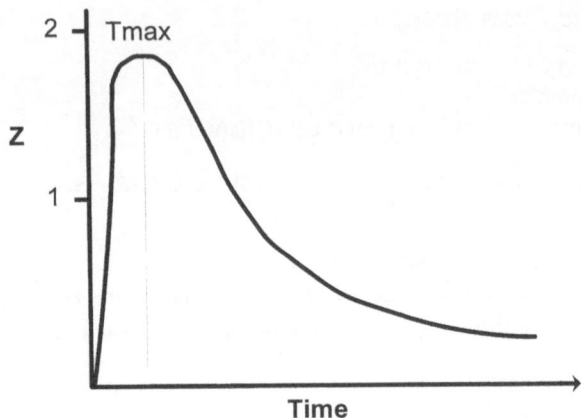

Fig. 1. Plot of functional hemolytic sites per cell (Z) against time. Convertase formation reaches a maximum at time *Tmax,* thereafter C2a decays from C4b and Z decreases.

cells expressing mutated or native forms of the protein provides a simple and rapid means for assessing structure/function activities of complement regulators *(11,12)*.

In classical pathway assays, the convertase-forming components are first incubated for the time required for maximum formation of C3 convertases on the cell surface, the *"Tmax" (13)*. *Tmax* is dependent on the amount of functionally active C4b on the cell surface, not C1 or C2, and should normally be less than 6 min. *Tmax* is determined by incubating EAC14b with C2 and measuring the number of functionally active sites per cell (Z) at specified times following addition of C2. A plot of Z against time demonstrates a curve with increasing Z values at early timepoints as convertases are formed to a maximum at the Tmax, followed by a decrease in Z as the convertases decay (**Fig. 1**).

3.1.1. Classical Pathway Hemolytic Assay (C4bp, DAF, CR1)

3.1.1.1. Determination of Time to Tmax:

1. Prepare EAC14b as described in Chapter 4, resuspend at 10^8 per mL in $DGVB^{2+}$. Prewarm to 30°C in a water bath.
2. Mix 2 mL EAC14b with 2 mL prewarmed C2 in $DGVB^{2+}$ (1 unit per mL; *see* **Note 1**) and incubate at 30°C in a shaking water bath.
3. Remove 200-μL aliquots at intervals from 1 to 15 min and add to 300 μL C^{rat}-EDTA (rat serum diluted 1:20 in GVB-40 m*M* EDTA). Incubate at 37°C for 1 h.
4. Add 2 mL cold saline to each sample at the end of the incubation, keep on ice until all incubations have been completed. Centrifuge at 1000*g* for 5 min and read absorbance of supernatant at 414 nm (A_{414}).

5. Calculate proportion of cells lysed (y) and hemolytic sites per cell (Z) at each time-point as follows:

$$y = (A_{414} \text{ sample} - A_{414} \text{ background}) / (A_{414} \text{ 100\%} - A_{414} \text{ background})$$

$$Z = -\ln (1 - y).$$

The background absorbance is the sum of the absorbances in supernatants from the cell blank (100 µL EAC14b + 400 µL DGVB^{2+}) and the serum color blank (200 µL DGVB^{2+}, 300 µL C-EDTA); the 100% absorbance is obtained from water-lysed cells (100 µL EAC14b, 100 µL DGVB^{2+}, 300 µL C-EDTA diluted in 2 mL water rather than saline).

6. Plot a graph of Z against time to assess *Tmax*.

3.1.1.2. ASSAY OF DECAY ACTIVITY

1. Prepare EAC14b as described in Chapter 4, resuspend at 10^8 per mL in DGVB^{2+}. Prewarm to 30°C in a water bath.
2. Mix 1 vol of EAC14b with 1 vol of prewarmed C2 in DGVB^{2+} and incubate at 30°C in a shaking water bath for time *Tmax*. The amount of C2 used is limited and should be previously determined to give 1–2 hemolytic sites remaining per cell following 15 min decay at 30°C with no inhibitor present (*see* Chapter 4).
3. Wash cells by centrifugation at 1000g for 5 min in ice-cold DGVB^{2+}, resuspend at 10^8 per mL in prewarmed (30°C) DGVB^{2+}.
4. Add an equal volume of cells to test sample (or buffer only as control; *see* **Note 2**) in DGVB^{2+}. Incubate at 30°C in a shaking water bath to enable convertases to decay.
5. Take 200-µL aliquots at suitable timepoints (such as 2, 4, 7, 10, 15, 20, 30, and 40 min) and add to 300 µL C^{rat}-EDTA (rat serum diluted 1:20 in GVB-40 mM EDTA). Incubate at 37°C for 1 h (*see* **Note 3**).
6. Add 2 mL cold saline to each sample at the end of the incubation, keep on ice until all incubations have been completed. Centrifuge at 1000g for 5 min and read absorbance of supernatant at 414 nm.
7. Calculate proportion of cells lysed (y) and hemolytic sites per cell (Z) remaining following decay period as described above.
8. Plot a graph of Z against time to assess increased rate of decay in presence of inhibitor (**Fig. 2**).

3.1.2. Alternative Pathway Hemolytic Assay (fH, DAF, CR1)

1. Prepare EAC4b3b as described in Chapter 4, resuspend at 10^8 per mL in DGVB^{2+}. Prewarm to 30°C in a water bath.
2. Incubate 1 vol EAC4b3b with 1 vol DGVB^{2+} containing excess P (3 µg), excess fD (100 ng), and a limiting amount of fB for 30 min at 30°C. The quantity of fB used should be previously determined to give 1–2 hemolytic sites remaining per cell following 15 min decay at 30°C.
3. Pellet EAC4b3bBbP cells by centrifugation at 1000g for 5 min. Wash in ice-cold GVB-EDTA and resuspend in prewarmed (30°C) GVB-EDTA at 10^8 per mL.

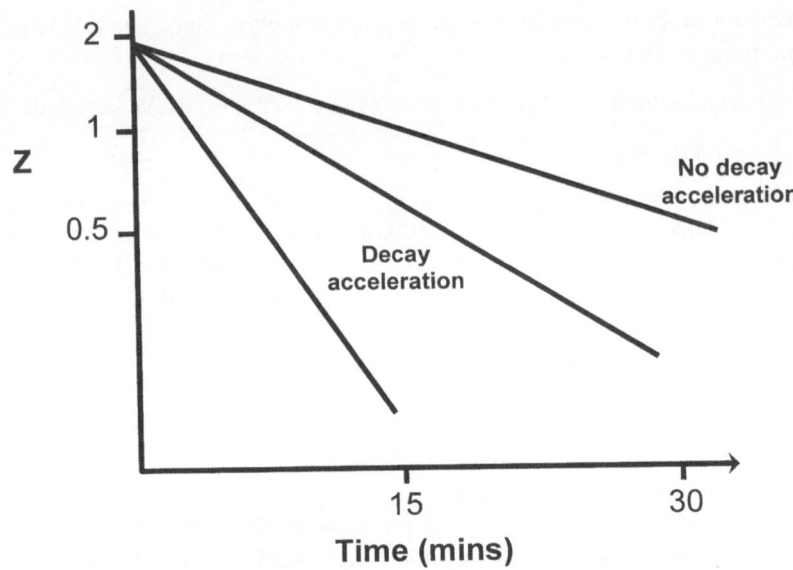

Fig. 2. Plot of functional hemolytic sites per cell (Z) against time of decay. Decay rate increases in the presence of decay accelerators.

4. Add an equal volume of cells to test sample (or buffer only as control; *see* **Note 2**) in GVB-EDTA. Incubate at 30°C in a shaking water bath to enable convertases to decay.
5. Take 200-μL aliquots at suitable time-points as above and add to 300 μL C^{rat}-EDTA (rat serum diluted 1:20 in GVB-40 mM EDTA). Incubate at 37°C for 1 h.
6. Add 2 mL cold saline to each sample at the end of the incubation, keep on ice until all incubations have been compl0eted. Centrifuge at 1000g for 5 min and read absorbance of supernatant at 414 nm.
7. Calculate proportion of cells lysed (y) and hemolytic sites per cell (Z) remaining following decay period as described in the previous section.
8. Plot a graph of Z against time to assess increased rate of decay in presence of inhibitor.

3.1.3. Assay of Decay-Accelerating Activity by Monitoring C3a Generation

1. Mix together purified components required for assembly of either CP or AP C3 convertase. Add dilutions of the test sample (or buffer as control) to the convertase-forming mixture in a total volume of 125 μL PBS (*see* **Note 4**). In either case the enzyme required to activate the convertase, fD or C1s, should be added to the mixture last and a sample should be taken just prior to addition to monitor C3a at time zero. Use the following amounts of reagents:
   a. Classical pathway: 0.2 μg C4; 2 μg C2; 8 μg C3; 0.1 μg C1s; 12 μL 0.1 M MgCl$_2$.

b. Alternative pathway: 0.05 µg C3i (for preparation of C3i, *see* **Note 5**); 2 µg fB; 10 µg C3; 0.2 µg fD; 12 µL 0.1 *M* MgCl$_2$.

2. Incubate at 37°C for 30 min prior to analysis for C3a generation by ELISA (Quidel) or radioimmunoassay (Amersham/Pharmacia).

3. Calculate% inhibition of C3a generation as follows:

$$\% \text{ inhibition} = 100 \times (\text{C3a}_{control} - \text{C3a}_{sample}) / \text{C3a}_{control}.$$

3.1.4. Assay of Decay Acceleration by Monitoring Inhibition of Total C Activity

1. Prepare cells as described in Chapter 4 for measurement of total C hemolytic activity. For an assessment of inhibition of CP activity, prepare sheep EA in GVB^{2+} at a final concentration of 10^8 per mL. For assessment of inhibition of AP activity prepare rabbit E in APB at 10^8 per mL.

2. Add 50 µL serum dilution to 50 µL of test sample or buffer only. Add 100 µL cells and incubate at 37°C for 30 min (*see* **Notes 6** and **7**).

3. Add 1 mL cold saline to each sample and centrifuge at 1000*g* for 5 min. Read absorbance of supernatant at 414 nm, calculate proportion of cells lysed (*y*) and hemolytic sites per cell (*Z*).

$$y = (A_{414} \text{ sample} - A_{414} \text{ background}) / (A_{414} 100\% - A_{414} \text{ background})$$

$$Z = -\ln(1 - y)$$

A_{414} background is the sum of the absorbance in supernatant from cell blank (100 µL E$_{sh}$A or E$_{rb}$ + 100 µL GVB^{2+} or APB) and serum color blank (100 µL GVB^{2+} or APB, 100 µL serum at appropriate dilution); A_{414} 100% is obtained by water lysis (100 µL E$_{sh}$A or E$_{rb}$ + 100 µL GVB^{2+} or APB, add 1 mL water rather than saline).

4. If serum concentration was set, plot graph of *Z* against concentration of inhibitor; if inhibitor concentration was set, plot *Z* against serum concentration.

3.2. Cofactor Activity

Cofactor activity is characterized by the ability of a protein to interact with either C3b or C4b, such that these molecules can be cleaved by fI to generate the inactive forms, iC3b, C3c, C3dg, C4d, and C4c. Cofactor activity is most easily demonstrated by incubating the cofactor and fI with substrate, C3b or C4b, and assessing protein cleavage by SDS-PAGE (**Fig. 3**) *(14–19)*. This approach can be used with either fluid phase cofactors, such as fH and C4bp, or cell-associated cofactors such as MCP and CR1. In the latter case, detergent lysates of cells expressing these proteins can be used directly in the assay *(20–22)*. As with decay-accelerating activity, the presence of a regulator on the cell surface protects the cell during C attack, decreasing C3b deposition and cell lysis. Measurement of cell-surface C3 fragment deposition and cytotoxicity is described in full in Chapters 5 and 10. As discussed above, these two techniques

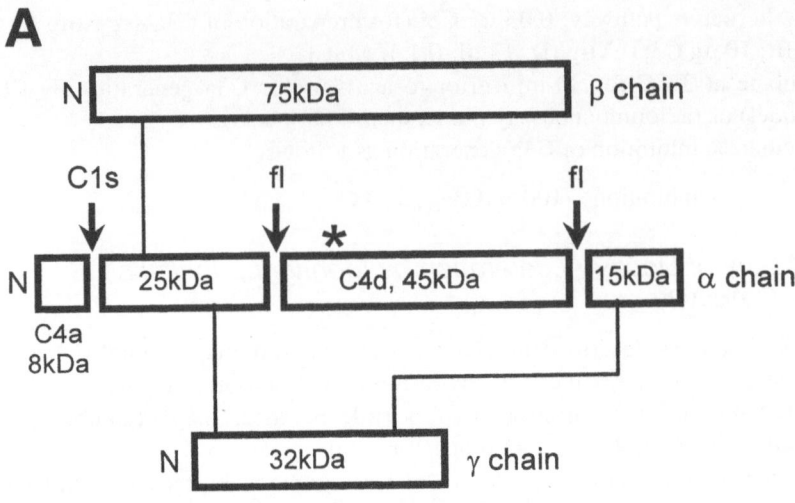

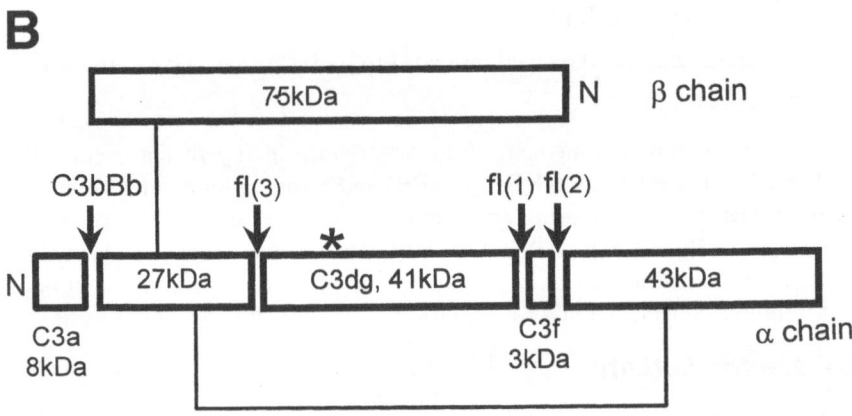

Fig. 3. Schematic representation of cleavage sites in (**A**) C4 and (**B**) C3. (**A**) C1s cleaves the α-chain of C4 resulting in formation of C4b and release of C4a. C4b is cleaved at two further sites by fI in the presence of C4bp, CR1, or MCP. This results in fragments derived from the α-chain of 25 kDa (33 kDa if C4$_{ma}$ is the substrate), 45 kDa (C4d), and 15 kDa. (**B**) The C3 convertase cleaves the α-chain of C3 resulting in formation of C3b and release of C3a. C3b is cleaved at two further sites by fI (fI$_{(1)}$ and fI$_{(2)}$) in the presence of fH, CR1, or MCP. This results in fragments derived from the α-chain of 68 kDa (76 kDa if C3$_{ma}$ is the substrate) and 43 kDa. In the presence of CR1 the α-chain can be cleaved again (fI$_{(3)}$) splitting the 68 kDa fragment into two fragments of 41 kDa and 27 kDa (35 kDa if C3$_{ma}$ is the substrate). The thioester (covalent binding site) is indicated by *, enzyme cleavage sites by arrows, and disulphide bonds by thin black lines. *N* indicates amino terminal end of peptide chain.

do not distinguish cofactor activity from other other forms of regulation, but are useful methods for assessing the activity of recombinant proteins on the surface of transfected cells, or activity of native proteins on cells with or without function-blocking antibodies.

3.2.1. Assay for Cofactor Activity

1. Native C3 and C4 are not substrates for fI. It is necessary first to generate C3b or C4b (*see* **Note 8**) or to hydrolyze the internal thioester by freeze/thaw cycles (*see* **Note 5**) or by treatment with a nucleophile such as methylamine. Methylamine-inactivated C3 or C4 (C3$_{ma}$, C4$_{ma}$) is prepared by incubating the protein (0.5 mg/mL) with 0.1 M methylamine for 2 h at 37°C in borate-buffered saline pH 8.0. The inactivated protein can then be dialyzed into PBS for the cofactor assay. If desired, C3b/C4b deposited on a cell surface can be used in place of purified components as substrate for fI when testing fluid phase inhibitors (*see* **Note 9**) *(3)*.

2. Label C3$_{ma}$/C4$_{ma}$ with ^{125}I to enable subsequent detection of α-chain fragments and quantitation of cleavage. Add 200 µL 0.1 mg/mL Iodogen in chloroform to a 12 × 75-mm glass tube, evaporate chloroform in a fume hood. Rinse tube with 200 µL PBS and add C3$_{ma}$ or C4$_{ma}$ (200 µL at 1 mg/mL in PBS) plus 1 mCi Na^{125}I (Amersham). Incubate on ice for 20 min, then separate labeled protein from free ^{125}I by gel filtration on Sephadex G25 (PD10 columns from Pharmacia are useful for this purpose). If desired, nonisotopic labeling methods can be used to detect cleavage of the α-chain (*see* **Note 10**).

3. Mix ^{125}I-labeled C3$_{ma}$ or C4$_{ma}$ (5–50 µg/mL final concentration) with fI (5–20% w/w) and test sample in a low ionic strength buffer such as PBS/3. Conditions of low ionic strength enhance protein–protein interactions and promote cleavage reactions. Soluble cofactors such as fH, C4bp or soluble recombinant forms of MCP and CR1, are typically used at 10–100% (w/w) the concentration of C3$_{ma}$/C4$_{ma}$. If purified cofactor (MCP or CR1) is not available, lysates of cells expressing the cofactor can be used directly in the assay (*see* **Note 11**); detergent should be kept in the incubation mix to preserve the activity of the protein *(17,23)*. The nonionic detergent NP40 (0.05%) is nondenaturing and preserves activity of proteins. If the cell lysate will form greater than 15% of the total reaction volume, lyse the cells in low ionic strength buffer (PBS/3).

    Include the following controls in the assay:
    a. C3$_{ma}$/C4$_{ma}$ only.
    b. C3$_{ma}$/C4$_{ma}$ plus fI, no cofactor.
    c. C3$_{ma}$/C4$_{ma}$ plus cofactor, no fI.

4. Incubate for 2 h at 37°C or until the desired degree of cleavage is obtained. The cleavage pattern obtained with fH-mediated cleavage of C3b is illustrated in **Fig. 4.** If the substrate is C4$_{ma}$ rather than C3$_{ma}$, it may be necessary to incubate for longer, up to 16 h. Stop the cleavage reaction by adding a portion of the incubation mix to an equal volume of reducing sodium dodecyl sulfate-polyacrylamide gel electrophoresis (SDS-PAGE) loading buffer (0.1 M Tris/HCl pH 6.8, 2%

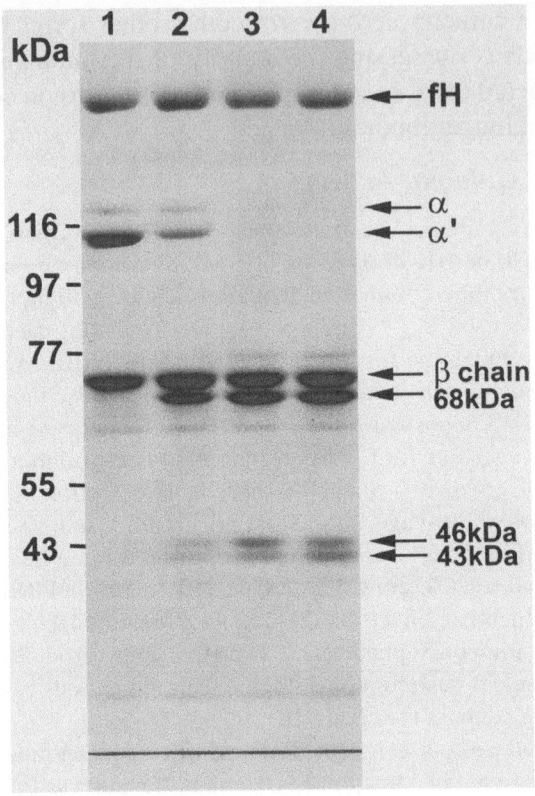

Fig. 4. Cleavage of C3b by fI in the presence of fH. This Coomassie-stained reducing SDS-PAGE gel illustrates the typical cleavage pattern obtained by fI-mediated cleavage of C3b with fH as cofactor, all components were purified from human plasma. In this case, C3 was cleaved to C3b by preincubation with a small amount of preformed CVFBb. The 68-kDa and 46-kDa fragments are the result of the first fI cleavage (fI$_{(1)}$ in **Fig. 3**), the 43-kDa fragment results from the second cleavage (fI$_{(2)}$). In the presence of CR1, the 68-kDa fragment would be further degraded to two fragments of sizes 41 kDa and 27 kDa. Track 1 illustrates the C3 banding pattern in the absence of fI. The uncleaved α-chain probably represents C3i (hydrolyzed C3) present in the C3 preparation, the fragment resulting from the first fI-mediated cleavage (fI$_{(1)}$) of C3i is probably that seen in lane 4 running with a slightly lower mobility than the β-chain. Lanes 1–3 represent aliquots of the incubation mixture sampled at increasing time-points following addition of fI.

SDS, 5% (v/v) β2-mercaptoethanol, 10% (v/v) glycerol, 0.05% (w/v) bromophenol blue). Boil sample for 2 min and load onto a 10% SDS-PAGE gel.
5. Following electrophoresis, quantitate radioactivity remaining in the intact α-chain and its cleavage fragments by either slicing the gel and measuring activity in a

γ-counter, or by analysis of the intact gel with a phosphoimager (*see* **Note 12**). Calculate % degradation of the α-chain.

3.3. Inhibition of MAC Formation

Regulators that function late in the C cascade include the membrane-associated regulators CD59 and HRF and the fluid phase regulators, S-protein and clusterin. These all function at a stage in the terminal pathway subsequent to formation of the C5b6 complex, hence their activities can be assessed using a "reactive lysis" system *(24,25)*. This relies on generation of the C5b6 complex, which can loosely associate with the cell membrane; addition of C7 creates C5b-7, which binds tightly to the membrane initiating C activation at this late stage in the cascade. The stable C5b6 complex is easily formed by C activation in acute phase serum, or serum that has been depleted of C7, and can be purified using classical chromatographic techniques. Reactive lysis can be initiated either by using purified components C5b6, C7, C8, and C9 or by incubation of serum with CVF (from *Naja naja kaouthia*) *(26,27)*. In the latter case, a convertase is formed that is not regulated by the human C-inhibitors, but is capable of cleaving C5 to C5b with consequent formation of C5b6.

Use of serum deficient in one of the terminal C components enables generation of cellular intermediates with "part-formed" terminal complexes (*see* Chapter 4). For example, incubation of cells with CVF and a serum deficient in (or depleted of) C8 results in formation of C5b-7 complexes on the cell surface. Similarly, incubation with serum deficient in C9 results in formation of cells bearing C5b-8 complexes. In both these cases, the cells remain intact (although cells bearing C5b-8 are rather "fragile"). MAC formation can be completed by use of purified components (C8 and/or C9) or C-EDTA. Sequential assembly of the MAC enables the investigator to "pin-point" the site of interaction of a regulator with the C proteins. Depletion of specific components from serum and generation of cellular intermediates is described in full in Chapter 4. The well-characterized MAC regulator, CD59, possesses a glycolipid anchor, enabling its incorporation into erythrocyte membranes by incubation at 37°C. When using a hemolytic assay to assess functional activity of human CD59, the source of cell is important. Human CD59 incorporated into the membrane of GPE is functional and protects against lysis. In contrast, protection is difficult to demonstrate following incorporation into some other species erythrocytes, notably sheep, because of a "masking" of the activity of the incorporated protein by the functional activity of CD59 native to the cell *(27)*. The assays described here are based on reactive lysis-mediated killing of GPE, they can be used to assess activity of incorporated regulators, such as CD59, and soluble regulators, such as S-protein and clusterin. S-protein and clusterin function by binding fluid phase C5b-7, hence C5b-7 and C5b-8 site assays cannot be used to assess function of these regulators. As is the case with regulators of the activation pathways, it is also possible to assay CD59 function on nucleated cells expressing recombinant forms of the protein, or activity of native proteins on cells with or without function-blocking antibodies. Expression of functional CD59 results in protection from C attack and a consequent decrease in cell cytotoxicity. Measurement of cytotoxicity is described in Chapter 5.

3.3.1. Measurement of MAC Inhibition in a Reactive Lysis System

1. Wash GPE three times in APB by centrifugation at $1000g$ for 5 min, resuspend at 2×10^8/mL (2% v/v) in APB with 10 µg/mL C5b6 and incubate at 37°C for 5 min (*see* **Note 13**).
2. Add C7 to a final concentration of 1 µg/mL, incubate for 15 min at 37°C.
3. Wash EC5b-7 cells twice in APB, resuspend at 2×10^8 per mL.
4. Titrate C8 to find a dose which, in the presence of an excess of C9, results in about 80% lysis by adding 100 µL APB containing C8 (at final concentrations between 1 ng/mL and 1 µg/mL) and C9 at 1 µg/mL to 100 µL EC5b-7. Incubate at 37°C for 30 min. Add 1 mL cold saline to each sample and centrifuge at $1000g$ for 5 min. Read absorbance of supernatant at 414 nm and calculate % lysis.

   % lysis $=100$ x $(A_{414}$ sample $- A_{414}$ cell blank$) / (A_{414}$ 100% $- A_{414}$ cell blank$)$

   The cell blank is absorbance in supernatant from 100 µL EC5b-7 incubated with 100 µL APB; 100% lysis is absorbance from 100 µL EC5b-7 + 100 µL APB diluted in 1 mL water.
5. Using this dose of C8, repeat the experiment with the test samples. Mix 100 µL EC5b-7 with 100 µL APB containing C8 (predetermined concentration), C9 (1 µg/mL) and test sample (*see* **Note 14**). Incubate at 37°C for 30 min. Calculate proportion of cells lysed (y) and hemolytic sites per cell (Z). Plot Z against concentration of inhibitor.

   $y = (A_{414}$ sample $- A_{414}$ cell blank$) / (A_{414}$ 100% $- A_{414}$ cell blank$)$

   $Z = -\ln (1 - y)$

3.3.2. CVF-Mediated "Reactive Lysis"

1. Wash GPE three times in APB by centrifugation at $1000g$ for 5 min, resuspend at 2×10^8 per mL in APB (*see* **Note 13**).
2. Determine amount of serum required to give approximately 80% lysis by mixing 100 µL GPE with 100 µL CVF (5 µg/mL) and 100 µL serum dilutions in APB. Incubate for 30 min at 37°C. Add 2 mL cold saline to each sample and centrifuge at $1000g$ for 5 min. Read absorbance of supernatant at 414 nm and calculate % lysis.

   % lysis $=100 \times (A_{414}$ sample $- A_{414}$ background$) / (A_{414}$ 100% $- A_{414}$ background$)$

Background absorbance is the sum of absorbance in supernatant from cell blank (100 μL GPE + 200 μL APB) and color blank (200 μL APB + 100 μL serum dilution); 100% is absorbance of supernatant from 100 μL GPE + 200 μL APB diluted in 2 mL water.

3. Using this amount of serum, repeat the experiment with dilutions of test sample or with cells in which CD59, or other GPI-anchored molecules, have been incorporated as described in **Note 14**. Calculate proportion of cells lysed (y) and hemolytic sites per cell (Z). Plot Z against concentration of inhibitor.

$$y = (A_{414} \text{ sample} - A_{414} \text{ background}) / (A_{414} \text{ 100\%} - A_{414} \text{ background})$$

$$Z = -\ln (1 - y)$$

3.3.3. C5b-7 and C5b-8 Site Assays

1. These assays follow essentially the same procedure as that described above in **Subheading 3.3.2.** The C8- or C9-depleted serum is first titrated on GPE to find a dilution at which there is no background lysis of the cells. Incubate 100 μL GPE with 100 μL CVF (5 μg/mL) and 100 μL serum dilution in APB for 30 min at 37°C. It should be possible to use depleted sera at a final dilution of 1:4 without any specific cell lysis.

2. Prepare GPE bearing either C5b-7 or C5b-8 sites by incubation of GPE and CVF with the appropriate dilution of either C8-depleted or C9-depleted serum respectively. Wash the cells in PBS/EDTA and resuspend at 2×10^8 per mL.

3. Determine the concentration of purified C8 and/or C9, or C-EDTA (used as a source of C8 and C9) required to give approximately 80% lysis; follow the procedure described above in **Subheading 3.3.2.**

4. Having determined the concentration of serum required to give approximately 80% lysis, repeat the experiment with dilutions of test sample or with cells in which CD59, or other GPI-anchored molecules, have been incorporated as described in **Note 14**. Calculate proportion of cells lysed (y) and hemolytic sites per cell (Z) as described above.

4. Notes

1. One unit of C2 is the amount required to give 63% lysis (the equivalent of one functional molecule offered per cell) in a hemolytic assay specific for C2.

2. It will be necessary to determine the appropriate concentration of inhibitor to use in these assays. Initially set up the test sample at various dilutions (with a 10-fold difference in concentration, for example). For the "natural" fluid phase inhibitors, fH and C4bp, concentration of inhibitor used in these assays is typically between 0.1 and 3 μg/mL. The concentration of purified DAF or CR1 used is typically in the region of 100 ng/mL.

3. The assay described here assesses the rate of decay of the convertase with a set concentration of inhibitor. It is equally possible to set the decay period (for example, to 15 min) and vary the concentration of inhibitor.

4. If the inhibitor has been purified from cell membranes (for example, DAF or CR1), include detergent in the mixture (0.1% NP40) to maintain solubility of the regulator.

5. C3i, also termed C3(H$_2$O), is prepared by freeze/thawing native C3. C3i can be separated from native C3 by anion exchange chromatography (DEAE Sepharose, Q Sepharose, monoQ), the hydrolyzed form elutes earlier in the NaCl gradient. Alternatively, C3 can be inactivated by incubation with nucleophiles such as methylamine (*see* **Subheading 3.2.1.**)

6. If the experimental variable is to be concentration of inhibitor, first titrate serum (in absence of inhibitor) and determine dilution that yields around 80% lysis. In subsequent experiments, use this dilution of serum and titrate test sample.

7. DAF purified from cell membranes possesses a glycolipid anchor, and will spontaneously incorporate into erythrocyte membranes *(8)*. This property can be exploited to prepare cells bearing DAF on their surface. Incorporation is achieved by incubating E (10^8 per mL) for 30 min at 37°C with inhibitor preparation in PBS/0.05% CHAPS. E are then washed twice in buffer and resuspended at 10^8 per mL.

8. Cleaved forms of C3 and C4 can be generated by limited trypsin cleavage *(28,29)*, or by cleavage with the C3 convertases, C3bBb, CVFBb, or C4b2a, generated using purified components in a fluid phase reaction. Cleaved fragments are easily separated from the convertase components by anion exchange chromatography.

9. Incubate EAC14b (10^8 per mL, bearing radiolabeled C4b) or EAC4b3b (10^8 per mL, bearing radiolabeled C3b) with cofactor (4 µg/mL) and fI (60 ng/mL) at 37°C for 1 h. Solubilize cells in SDS-PAGE loading buffer and assess cleavage of α' chain by SDS-PAGE as described in method 3.2.1.

10. If the cofactor is purified, cleavage can be demonstrated simply by Coomassie or silver stain of the gel. Alternatively, label C3$_{ma}$ or C4$_{ma}$ with biotin and detect cleavage fragments following Western blotting by incubation with HRPO-conjugated streptavidin and development by enhanced chemiluminesence *(18)*. Immunoblotting using polyclonal anti-C3 or C4 can also be used to detect cleavage fragments *(21)*.

11. Harvest cells and wash several times in PBS. Count cells and solubilize in PBS, 1% NP40, 10 m*M* EDTA, 1 m*M* PMSF, 1 µg/mL pepstatin A, 1 µg/mL leupeptin for 30 min on ice. Solubilize 8×10^7 cells in 1 mL buffer, unless working with erythrocytes in which case solubilize 5×10^9 hypotonically lysed cells in 1 mL solubilization buffer. Spin cell lysate at 30,000g for 15 min at 4°C. Harvest supernatant for use in cofactor assay.

12. If the β-chain cannot be distinguished from fragments of the α-chain because of the presence of C3a in C3$_{ma}$, use C3b rather than C3$_{ma}$, the mobilities will then be distinct; *see* **Fig. 3**.

13. EA can be used to assay for MAC-inhibitory activity, although be aware that human CD59 activity is difficult to demonstrate using this system. If using EA, incubate 100 µL EA (10^8 per mL) with 100 µL GVB and 100 µL serum dilution

for 30 min at 37°C. Determine dilution of serum giving approx 80% lysis, and use this serum dilution to demonstrate inhibitory activity in test samples.

14. Incorporate GPI-anchored proteins by incubating GPE (10^8 per mL) for 30 min at 37°C with inhibitor preparation in PBS/0.05% CHAPS. E are then washed twice in buffer and resuspended at 10^9 per mL. CD59 can also be incorporated into cells bearing C5b-7 or C5b-8 sites, enabling demonstration of its site of action.

References

1. Pensky, J., Levy, L., and Lepow, I. (1961) Partial purification of a serum inhibitor of C1'esterase. *Biol, J. Chem.* **236,** 1674.
2. Whaley, K. and Ruddy, S. (1976) Modulation of the alternative complement pathways by beta 1 H globulin. *J. Experiment. Med.* **144,** 1147.
3. Gigli, I., Fujita, T., and Nussenzweig, V. (1979) Modulation of the classical pathway C3 convertase by the plasma proteins C4 binding protein and C3b inactivator. *Proc. Natl. Acad. Sci. USA* **76,** 6596.
4. Morgan, B. P. and Meri, S. (1994) Membrane proteins that protect against complement lysis. *Springer Semin. Immunopathol.* **15,** 369.
5. Weiler, J. M., Daha, M. R., Austen, K. F., and Fearon, D. T. (1976) Control of the amplification convertase of complement by the plasma protein beta1H. *Proc. Natl. Acad. Sci. USA* **73,** 3268.
6. Fearon, D. T. (1979) Regulation of the amplification C3 convertase of human complement by an inhibitory protein isolated from human erythrocyte membranes. *Proc. Natl. Acad. Sci. USA* **76,** 5867.
7. Nicholson-Weller, A., Burge, J., and Fearon, D. T. (1982) Isolation of a human erythrocyte membrane glycoprotein with decay-accelerating activity for C3 convertases of the complement system. *J. Immunol.* **129,** 184.
8. Medof, M. E., Kinoshita, T., and Nussenzweig, V. (1984) Inhibition of complement activation on the surface of cells after incorporation of decay-accelerating factor (DAF) into their membranes. *J. Experiment. Med.* **160,** 1558.
9. Seya, T., Holers, V. M., and Atkinson, J. P. (1985) Purification and functional analysis of the polymorphic variants of the C3b/C4b receptor (CR1) and comparison with H, C4b-binding protein (C4bp), and decay accelerating factor (DAF). *J. Immunol.* **135,** 2661.
10. Brodbeck, W. G., Liu, D., Sperry, J., Mold, C., and Medof, M. E. (1996) Localization of classical and alternative pathway regulatory activity within the decay-accelerating factor. *J. Immunol.* **156,** 2528.
11. Lublin, D. M. and Coyne, K. E. (1991) Phospholipid-anchored and transmembrane versions of either decay-accelerating factor or membrane cofactor protein show equal efficiency in protection from complement-mediated cell damage. *J. Experiment. Med.* **174,** 35.
12. Oglesby, T. J., Allen, C. J., Liszewski, M. K., White, D. J., and Atkinson, J. P. (1992) Membrane cofactor protein (CD46) protects cells from complement-mediated attack by an intrinsic mechanism. *J. Experiment. Med.* **175,** 1547.

13. Whaley, K. (1985) Measurement of complement, in *Methods in Complement for Clinical Immunologists* (Whaley, K., ed.), Churchill Livingstone, Edinburgh, p. 77.

14. Fujita, T. and Nussenzweig, V. (1979) The role of C4-binding protein and beta 1H in proteolysis of C4b and C3b. *J. Experiment. Med.* **150,** 267.

15. Yoon, S. H. and Fearon, D. T. (1985) Characterization of a soluble form of the C3b/C4b receptor (CR1) in human plasma. *J. Immunol.* **134,** 3332.

16. Seya, T., Turner, J. R., and Atkinson, J. P. (1986) Purification and characterization of a membrane protein (gp45-70) that is a cofactor for cleavage of C3b and C4b. *J. Experiment. Med.* **163,** 837.

17. Seya, T. and Atkinson, J. P. (1989) Functional properties of membrane cofactor protein of complement. *Biochem. J.* **264,** 581.

18. Gordon, D. L., Kaufman, R. M., Blackmore, T. K., Kwong, J., and Lublin, D. M. (1995) Identification of complement regulatory domains in human factor H. *J. Immunol.* **155,** 348.

19. Seya, T., Nakamura, K., Masaki, T., C. Ichihara-Itoh, Matsumoto, M., and Nagasawa, S. (1995) Human factor H and C4b-binding protein serve as factor I-cofactors both encompassing inactivation of C3b and C4b. *Molec. Immunol.* **32,** 355.

20. Adams, E. M., Brown, M. C., Nunge, M., Krych, M., and Atkinson, J. P. (1991) Contribution of the repeating domains of membrane cofactor protein (CD46) of the complement system to ligand binding and cofactor activity. *J. Immunol.* **147,** 3005.

21. van den Berg, C. W., J. M. Perez de la Lastra, Llanes, D., and Morgan, B. P. (1997) Purification and characterization of the pig analogue of human membrane cofactor protein (CD46/MCP). *J. Immunol.* **158,** 1703.

22. Tsujimura, A., Shida, K., Kitamura, M., Nomura, M., Takeda, J., Tanaka, H., Matsumoto, M., et al. (1998) Molecular cloning of a murine homologue of membrane cofactor protein (CD46): Preferential expression in testicular germ cells. *Biochem. J.* **330,** 163.

23. Seya, T., Hara, T., Iwata, K., Kuriyama, S., Hasegawa, T., Nagase, Y., et al. (1995) Purification and functional properties of soluble forms of membrane cofactor protein (CD46) of complement: identification of forms increased in cancer patients' sera. *Int. Immunol.* **7,** 727.

24. Lachmann, P. J. and Thompson, R. A. (1970) Reactive lysis: the complement-mediated lysis of unsensitized cells. II. The characterization of activated reactor as C56 and the participation of C8 and C9. *J. Experiment. Med.* **131,** 643.

25. Thompson, R. A. and Lachmann, P. J. (1970) Reactive lysis: the complement-mediated lysis of unsensitized cells. I. The characterization of the indicator factor and its identification as C7. *J. Experiment. Med.* **131,** 629.

26. Whitlow, M. B., Iida, K., Stefanova, I., Bernard, A., and Nussenzweig, V. (1990) H19, a surface membrane molecule involved in T-cell activation, inhibits channel formation by human complement. *Cell. Immunol.* **126,** 176.

27. van den Berg, C. W. and Morgan, B. P. (1994) Complement-inhibiting activities of human CD59 and analogues from rat, sheep, and pig are not homologously restricted. *J. Immunol.* **152,** 4095.

28. Bokisch, V. A., Muller-Eberhard, H. J., and Cochrane, C. G. (1969) Isolation of a fragment (C3a) of the third component of human complement containing anaphylatoxin and chemotactic activity and description of an anaphylatoxin inactivator of human serum. *J. Experiment. Med.* **129,** 1109.

29. Avery, V. M. and Gordon, D. L. (1993) Characterization of factor H binding to human polymorphonuclear leukocytes. *J. Immunol.* **151,** 5545.

7

Immunochemical Measurement of Complement Components and Activation Products

Reinhard Würzner

1. Introduction

Activation of the complement system by the classical, mannan binding lectic (MBL) or alternative pathway results in the generation of multiple complement proteolytic cleavage products, recruited from these inactive precursor molecules in a sequentially proceeding cascade (Chapter 1). This leads to an alteration of their antigenic pattern and thus to the consumption of the native molecule. Concomitantly, complement activation initiates formation of multimolecular complexes such as the C3 or C5 convertases or the terminal complement complex (TCC). The latter can be generated on biological membranes or in the fluid phase (C5b-9). Activation-dependent changes on molecules or complexes can be revealed by the disappearance of native-restricted epitopes, and by the appearance of neoepitopes (*1*), and are usually assessed by means of monoclonal antibodies in enzyme-linked immunosorbent acids (ELISAs). This chapter focuses on the sensitive and specific quantitation of (native) complement proteins to evaluate the inactivated (background) status, and also on that of their fragments and complexes to reliably assess complement activation in vivo, with particular respect to neoepitopes of C9 appearing only when C9 is incorporated into the terminal complement complex.

1.1. Assays Other than ELISA to Quantitate Complement Components or Activation Products

Although the majority of these assays are now becoming less common in research laboratories, they still have their place in routine complement diagnostics. They comprise methods assessing the presence and integrity of a

From: *Methods in Molecular Biology, vol. 150: Complement Methods and Protocols*
Edited by: B. P. Morgan © Humana Press Inc., Totowa, NJ

protein, like nephelometric assays, radial immunodiffusion (Mancini), or electroimmunodiffusion (rocket electrophoresis, Laurell) techniques *(2–5)*. Functional methods assessing the hemolytic capacity of the entire system, its pathways, or even of single components are described in Chapter 4.

Identification of complement activation in tissues or on blood cells by immunofluorescence are described in Chapter 10. They may also be optimized by using native-restricted or neoepitope-specific MAbs to discriminate local complement activation in situ from passive trapping of (native) complement components.

1.2. Immunochemical Assays Using Native Restricted Monoclonal Antibodies

Isolated immunochemical determination of concentrations of individual (native) components is often of limited value in clinical practice as these levels exhibit wide normal ranges (both interindividually and intraindividually) that obscure minor deviations. This is mainly because of different rates of synthesis and degradation and renal or hepatic clearance or because of multiple interactions with other serum proteins or cellular receptors *(6)*. The acute phase behavior of most of the complement components will further cloud any changes.

However, these assays are particularly useful for assessing deficiency states, as detailed in Chapter 11, where they allow sensitive discrimination between subtotal and complete deficiency. However, it has to be born in mind that it is wrong to assume that approximately half normal concentrations indicate heterozygous deficiency—even obligate heterozygous subjects, e.g., parents or children of complement deficient subjects, sometimes present with almost normal concentrations of the component in question. Furthermore, these assays are particularly suitable for assessing local biosynthesis of complement components.

The optimal design, as outlined in detail elsewhere *(7)*, uses the most specific, usually native-restricted MAb as coating antibody. This ensures that the epitope in question is bound to the solid phase via the coating MAb and is thus in a less denatured form, when compared to the solid-phase adsorbed status, and avoids saturation of the solid phase with irrelevant fragments. MAbs have several advantages over polyclonal antibodies: (1) once the stock is exhausted, an identical antibody can be still produced and (2) the specificity is narrow and thus high. Most importantly, native restricted polyclonal antibodies directed against complement proteins have not been described yet.

The sample to be assayed is added in an appropriate dilution and measured by means of a second antibody. This second antibody is preferably directly labeled with the enzyme as it reduces crossreactions by avoiding antispecies antibodies in the next step, which may react with the solid phase MAb. How-

Table 1
Native-Restricted and Neoepitope-Specific
MAbs Directed Against Complement Components

		NR	NEO
C1	C1q, C1s	√	√
C4	iC4, C4b, iC4b, C4c, C4d		√
C2		√	
C3	iC3, C3b, iC3b, C3c, C3dg, C3d, C3g, C3a, C3a-desArg		√
B	Ba, Bb		√
C5	C5a, C5a-desArg	√	√
C6		√	
C9	C5b-9(m), SC5b-9, poly C9	√	√

Further details and references of MAbs directed against native-restricted epitopes (NR) or neoepitopes (NEO) are detailed elsewhere *(7)* or in **Table 2** (for anti-C9 neoepitopes only).

ever, the detection antibody itself may already crossreact with the coating antibody if it is not derived from the same species. Thus, a good MAb is preferable, although polyclonal antibodies may be used as well at this step, as the specificity of the assay is determined by the coating antibody.

Native specific MAbs have been characterized for many complement proteins (**Table 1**), also reviewed elsewhere *(7)*. In addition to their use for reliable quantitation of the unactivated molecule, they have been used to inhibit complement activation in vitro and thus represent potential therapeutic agents for in vivo interventions *(8)*.

1.3. Immunochemical Assays
Using Neoepitope-Specific Monoclonal Antibodies

Multiple studies have shown that the decrease in concentration of the uncleaved component is less sensitive for assessing complement activation than the increase in cleavage products of the complement activation or the complexes containing that particular component. This is easily understood: a rise of a particular concentration from 1 to 5% is much easier to detect (fivefold increase!) than a decrease from 99 to 95% (which is within the error of the assay). Nevertheless, an accurate assessment of complement activation in vivo requires the simultaneous determination of both native and activated complement proteins: low amounts of native proteins in the first place cannot generate as much activation product as high concentrations. Most quantitations, however, are hampered by the fact that the majority of antibodies recognize both native and activated forms and thus are not specific for the latter. Furthermore, the removal of the native form by immunochemical or functional means such as immunoprecipitation methods is error-prone as it can never be complete.

Table 2
Neoepitope-Specific MAbs Directed Against C9

Antibody	Reference
Poly C9-MA	*(9)*
aE11/M0777	*(10)*, Dako
bC5	*(10)*
3B1, 3D8, 2F3, 1A12	*(11)*
056B-75/A209	*(12)*, Quidel
1B4	*(13)*
WU 7-2, WU 13-15	*(14)*
X-11	*(15)*, Biomedicals
B7	*(16)*

Hence, a sensitive assay for the assessment of activated complement molecules may additionally measure 5–10% of contaminating native molecules if these have not been depleted in the preceding step. This is the reason why MAbs specific for activation-dependent epitopes (neoepitopes) are preferred in order to distinguish reliably between activated and native states of complement proteins *(1,7)*.

Neoepitope-specific MAbs have been characterized for many complement proteins (**Table 1**), also reviewed elsewhere *(7)*. Assays based on these MAbs have shown to reliably measure complement activation in vivo. They have been used to follow the course of a disease, to reveal exacerbations, and to evaluate the success of a treatment. In particular, these assays have been used to assess the biocompatibility of extracorporeal membranes or to evaluate the therapeutic use of inhibitory antibodies *(8)*.

1.4. Immunochemical Assays Using Anti-C9 Neoepitope-Specific Monoclonal Antibodies

The binding regions of all neoepitope-specific anti-TCC MAbs described so far (**Table 2**) are located on the C9 moiety of the TCC, with the exception of one MAb (aE11), which also crossreacts with C8α within the nascent complex *(17)*, which is likely to be a reflection of the fact that C8α is structurally and functionally related to C9. The predominant presence of the neoepitopes on C9 mirrors the marked conformational changes, which C9 undergoes during TCC assembly, but may also be caused by the fact that it is the most abundant molecule in the TCC.

Despite the fact that the membrane-integrated form (C5b-9(m)) is structurally different from the fluid phase form (SC5b-9), especially with respect to the number of C9 molecules per complex, no MAb has been described so far that is able to distinguish between the two forms. Because vitronectin (S-protein) is also able to integrate into the membrane attack complex (MAC) after generation of C5b-9(m), this cannot be used for discrimination.

For poly C9, the number of C9 molecules present may also play a role in the appearance and disappearance of activation specific epitopes. Poly C9 can be made using zinc, generating much bigger complexes (with more C9 molecules) than by the method using magnesium. One MAb (X-11) failed to react to poly C9 generated by the latter method, whereas the former was not used in these experiments *(15)*. It is possible that MAbs will become available that can discriminate between C5b-9(m), SC5b-9, poly $C9_{(2\text{-few C9 molecules})}$, and poly $C9_{(\text{many C9 molecules})}$.

In any case, most of the other anti-TCC MAbs show differences, for example some precipitate C5b-9 complexes whereas others do not *(11)*. In addition, some crossreact with animal sera *(11)*, which is useful for animal studies, whereas others do not even crossreact with primate sera *(15)*.

Several sensitive ELISAs, based on these neoepitope-specific MAbs, which are able to detect TCC in EDTA-plasma, have been described *(12–14,18,19)*. The protocol for an ELISA procedure using one of these MAbs is detailed below.

2. Materials

2.1. Sandwich ELISA for the Determination of Fluid-Phase TCC (SC5b-9)

1. Spectrophotometer capable of reading optical densities in 96-well microtiter plates at 405 or 410 nm and 490 nm (reference filter).
2. Optional: automatic washer for 96-well microtiter plates.
3. Optional: multichannel pipet.
4. 96-well flat-bottom microtiter plates.
5. Appropriate laboratory gloves.
6. Phosphate-buffered saline (PBS).
7. Coating buffer consisting of 0.2 *M* sodium carbonate, pH 10.6.
8. Blocking buffer consisting of coating buffer or PBS, supplemented with 0.5–1% gelatine or bovine serum albumin.
9. Incubation and washing buffer (IWB) consisting of PBS supplemented with 0.05% Tween-20 (Sigma).
10. Standard: pool of normal human serum activated with baker's yeast (30 mg/mL plasma, incubate 2 h at 37°C, centrifuge for 10 min at 10,000*g*, contains about 300-600 µg TCC/mL).
11. Samples to be tested, sera should be supplemented with 20 m*M* ethylenediamine-tetraacetic acid (EDTA) (final concentration in IWB) to avoid further in vitro complement activation (*see* **Note 1**).
12. MAb WU 7–2 or WU 13–15 [coating and detection antibodies and a reference sample are available from the author upon request *(14)*; similar anti-TCC MAbs (listed in **Table 2**) can be obtained from other sources].

13. Polyclonal immunoglobulin to a terminal pathway component, e.g., goat anti-C6 IgG (any good anti-C6 or anti-C7 IgG may be used), biotinylated.
14. Streptavidin-horseradish peroxidase (Boehringer Mannheim, 1089152) or streptavidin-alkaline phosphatase (DAKO, D 0396).
15. Horseradish peroxidase development system: use 0.5–2 mM ABTS (2,2 azino-di(3-ethyl)-benzthiazoline sulphonate, e.g., (Boehringer Mannheim, 102946)) in 0.1 M acetate buffer (alternatively: supplemented with 0.05 M sodium phosphate), pH 4.0–4.2, to which H_2O_2 (2.5 mM, 30 µL of 3% H_2O_2 per 10 mL acetate buffer) is added immediately before use; or alkaline phosphatase system: dissolve 1 tablet pNPP (p-nitrophenyl phosphate, Sigma) in 5 mL substrate buffer, consisting of 0.1 M glycine, 1 mM $MgCl_2$, 1 mM $ZnCl_2$, pH 10.4.
16. Optional: 10% H_2SO_4 (for ABTS) or 3 N NaOH (for pNPP) as stop solution

3. Methods

1. Coat the wells of a standard microplate by overnight incubation at 4–8°C (works better than using shorter incubation times for higher incubation temperatures) with 100 µL of WU 7-2 or WU 13-15, 15 µg/mL in coating buffer (preferred). Do not coat the outer wells of the plate as they usually yield less-reliable results, even if the plate is always placed in a moist box as recommended for all steps. Seal the plate with parafilm (American National Can Co.) to reduce evaporation effects. The plates can be stored at 4°C for up to 1 wk.
2. Remove the coating buffer by deflecting the plate but do not wash with IWB as the detergent (Tween) will affect the following blocking step. To each well, add 150–200 µL blocking buffer in order to block the whole well and avoid nonspecific adsorption to those parts of the wells that become exposed when the plate is not held or placed exactly horizontally (e.g., when you carry the plate). Cover the plate in parafilm and block for about 30 min at room temperature.
3. Wash plate with IWB at room temperature, at least three times between all incubation steps, using 200–300 µL per well. Wash by hand using a multichannel pipet or a plastic bottle, or use an automatic washer. An experienced worker will be faster and just as reliable as an automatic washer. The latter is advantageous only when many plates are being used.
4. Use gloves from this step onward, as human samples are handled. Keep IWB and samples cold (on crushed ice). Apply sera, 1/20 diluted in IWB supplemented with EDTA, or plasma 1/2 diluted in IWB, or activated serum, 1/400 diluted in IWB, and standard activated serum in a doubling dilution series from 1/100 to 1/128,000 diluted in IWB and incubate for 1–2 h at 4°C. Longer incubation times, even overnight, are also possible but offer no added value. Incubation at 4°C is essential to reduce further complement activation, especially when the samples are not supplemented with EDTA (*see* **Note 1**). If the serum samples are to be tested virtually undiluted (e.g., when testing deficient subjects), add more Tween in a smaller volume of IWB.
5. Wash the plate three times as in **step 3**, but be particularly careful at the first wash, as some wells may contain very high concentrations of the antigen and

others very low. Under these circumstances, even a very short exposure of the latter well to higher amounts of antigen from a neighboring well may yield falsely positive results. In addition, the amount of potentially hazardous agents in the wells is largest at this step.

6. Apply detection antibody, e.g., goat anti-C6, biotinylated, diluted 20 μg/mL in IWB for about 1 h at room temperature.
7. Wash plate as in **step 3**.
8. Apply conjugate streptavidin-horseradish peroxidase or streptavidin-alkaline phosphatase, diluted according to the manufacturer (usually 1/1000 or higher) in IWB for about 1 h at room temperature.
9. Wash plate as in **step 3**.
10. Add substrate (ABTS or pNPP, 100 mL per well) rapidly and incubate at room temperature for several minutes until the upper standard reaches an optical density of approx 1.5. Add 100 μL of stop solution if immediate measurement is not possible.
11. Read in a spectrophotometer for 96-well microtiter plates at 405/410 nm using 490 nm as reference filter. For comparison it is recommended to measure at different time intervals to enable calculation of an increase in optical density per minute.

4. Notes

1. Collection and preservation of samples to be analyzed for complement activation products is complicated by the fact that complement undergoes a continuous low-grade activation both in vitro and in vivo. Thus, complement activation products are normally present in low amounts in plasma. In order to obtain reliable results that reflect the in vivo situation as closely as possible, it is crucially important to avoid in vitro activation of complement. There are three basic criteria for correct sample collection *(20)*. First, the sample should be collected into a tube containing EDTA (10–20 m*M* final concentration), which inhibits both the classical and the alternative pathways; after sampling, the EDTA has to be mixed with the blood by gentle shaking (the generation of air bubbles should be avoided!). Second, the sample temperature must be kept low throughout the whole procedure (immediate and "cold" centrifugation, storage on crushed ice when the sample is not placed in a cooled apparatus). Third, the time from collection to separation from blood cells and final storage of the plasma, preferably at –70°C, should be as short as possible. For storage the sample should be divided into several small aliquots to avoid freeze-thaw cycles. Storage at –20°C is possible for shorter periods but may already lead to antigenic changes and exposure of neoepitopes, at least for the rather labile components C3 and C4, after a few weeks. In contrast, TCC is rather stable in vitro and false positive results because of inappropriate storage are much less common.
2. There are numerous advantages to the use of anti-C9 neoepitope-specific MAbs. First, TCC has a much longer half-life in vivo than, e.g., C5a because of immediate binding of the latter to its cell-membrane receptor. Second, data from in vitro experiments suggest that terminal complement activation may also occur in the

absence of C3 activation or C5a generation, by cleavage of C5 caused by enzymes released from injured tissues or by oxygen radicals. Third, the stability of TCC is much higher than that of the early activation products both in vitro and in vivo, yielding more accurate data. In addition, anti-C9 neoepitope-specific ELISA are particularly useful in assessing whether terminal complement deficient subjects carry very low levels of functionally active proteins *(21–23)*. The amount of TCC is a measure for the whole system as it includes the assessment of all major components, including C3 (opsonization, chemotaxis), C5 (chemotaxis), and C5b-9 (cell signaling, MAC). However, it does not necessarily reflect an arrest (e.g., deficiency) in one of the pathways as one functionally active pathway is sufficient for TCC generation. The unique feature that the TCC-ELISA does not only prove the presence of all major components, but also their functional activity makes this immunochemical assay a very sensitive and specific functional test. Thus, anti-C9 neoepitope-specific MAbs are strongly recommended for the assessment of complement activation.

References

1. Mollnes, T. E. and Harboe, M. (1993) Neoepitope expression during complement activation—a model for detecting antigenic changes in proteins and activation of cascades. *Immunologist* **1**, 43–49.
2. Whaley, K. (1985) *Methods in Complement for Clinical Immunologists.* Churchill-Livingstone, Edinburgh.
3. Harrison, R. A. (1996) Purification, assay, and characterization of complement proteins from plasma, in *Handbook of Experimental Immunology* (Herzenberg, L. A., Weir, D. M., Herzenberg, L. A., and Blackwell, C., eds.), Blackwell, Cambridge, MA, pp. 75.1–75.50.
4. Kirschfink, M. (1997) The clinical laboratory: testing the complement system, in *The Complement System* (Rother, K., Till, G. O., and Hänsch, G. M., eds.), Springer, Berlin, pp. 522–547.
5. Whaley, K. and North, J. (1997) Haemolytic assays for whole complement activity and individual components, in *Complement: A Practical Approach* (Dodds, A. W. and Sim, R. B., eds.), IRL Press, Oxford, pp. 19–47.
6. Oppermann, M., Höpken, U., and Götze, O. (1992) Assessment of complement activation in vivo. *Immunopharmacology* **24**, 119–134.
7. Würzner, R., Mollnes, T. E., and Morgan, B. P. (1997) Immunochemical assays for complement components, in *Immunochemistry 2: A Practical Approach* (Johnstone, A. P. and Turner, M. W., eds.), Oxford University Press, Oxford, pp. 197–223.
8. Würzner, R. (1993) Monoclonal antibodies against the terminal complement components, in *Activators and Inhibitors of Complement* (Sim, R. B., ed.), Kluwer Academic, Doordrecht, The Netherlands, pp. 167–180.
9. Falk, R. J., Dalmasso, A. P., Kim, Y., Tsai, C. H., Scheinman, J. I., Gewurz, H., and Michael, A. F. (1983) Neoantigen of the polymerized ninth component of complement. Characterization of a monoclonal antibody and immunohistochemical localization in renal disease. *J. Clin. Invest.* **72**, 560–573.

10. Mollnes, T. E., Lea, T., Harboe, M., and Tschopp, J. (1985) Monoclonal antibodies recognizing a neoantigen of poly(C9) detect the human terminal complement complex in tissue and plasma. *Scand. J. Immunol.* **22,** 183–195.
11. Hugo, F., Jenne, D., and Bhakdi, S. (1985) Monoclonal antibodies against neoantigens of the terminal C5b-9 complex of human complement. *Biosci. Rep.* **5,** 649–658.
12. Kolb, W. P., Morrow, P. R., Jensen, F. C., and Tamerius, J. D. (1988) Development of a highly sensitive capture EIA for the quantification of the SC5b-9 complex in human plasma using a monoclonal antibody reactive with a poly-C9 neoantigenic determinant. *Complement* **5,** 213–214.
13. Kusunoki, Y., Takekoshi, Y., and Nagasawa, S. (1990) Using polymerized C9 to produce a monoclonal antibody against a neoantigen of the human terminal complement complex. *J. Pharmacobiodyn.* **13,** 454–460.
14. Würzner, R., Schulze, M., Happe, L., Franzke, A., Bieber, F. A., Oppermann, M., and Götze, O. (1991) Inhibition of terminal complement complex formation and cell lysis by monoclonal antibodies. *Compl. Inflamm.* **8,** 328–340.
15. Würzner, R., Xu, H., Franzke, A., Schulze, M., Peters, J. H., and Götze, O. (1991) Blood dendritic cells carry terminal complement complexes on their cell surface as detected by newly developed neoepitope specific monoclonal antibodies. *Immunology* **74,** 132–138.
16. Kemp, P. A., Spragg, J. H., Brown, J. C., Morgan, B. P., Gunn, C. A., and Taylor, P. W. (1992) Immunohistochemical determination of complement activation in joint tissues of patients with rheumatoid arthritis and osteoarthritis using neoantigen-specific monoclonal antibodies. *J. Clin. Lab. Immunol.* **37,** 147–162.
17. Tschopp, J. and Mollnes, T. E. (1986) Antigenic crossreactivity of the alpha subunit of complement component C8 with the cysteine-rich domain shared by complement component C9 and low density lipoprotein receptor. *Proc. Natl. Acad. Sci. USA* **83,** 4223–4227.
18. Mollnes, T. E., Lea, T., Froland, S. S., and Harboe, M. (1985) Quantification of the terminal complement complex in human plasma by an enzyme-linked immunosorbent assay based on monoclonal antibodies against a neoantigen of the complex. *Scand. J. Immunol.* **22,** 197–202.
19. Hugo, F., Krämer, S., and Bhakdi, S. (1987) Sensitive ELISA for quantitating the terminal membrane C5b-9 and fluid phase SC5b-9 complex of human complement. *J. Immunol. Meth.* **99,** 243–251.
20. Mollnes, T. E., Garred, P., and Bergseth, G. (1988) Effect of time, temperature and anticoagulants on in vitro complement activation: consequences for collection and preservation of samples to be examined for complement activation. *Clin. Exp. Immunol.* **73,** 484–488.
21. Würzner, R., Orren, A., Potter, P., Morgan, B. P., Ponard, D., Späth, P., et al. (1991) Functionally active complement proteins C6 and C7 detected in C6- or C7- deficient individuals. Clin. Exp. Immunol., **83,** 430–437.
22. Würzner, R., Platonov, A. E., Beloborodov, V. B., Pereverzev, A. I., Vershinina, I. V., Fernie, B. A., et al. (1996) How partial C7 deficiency with chronic and

recurrent bacterial infections can mimic total C7 deficiency: temporary restoration of host C7 level following plasma transfusion. *Immunology* **88,** 407–411.

23. Witzel-Schlömp, K., Hobart, M. J., Fernie, B. A., Orren, A., Würzner, R., Rittner, C., et al. (1998) Heterogeneity in the genetic basis of human complement C9 deficiency. *Immunogenetics* **48,** 144–147.

8

Complement Deposition in Tissues

Antti Väkevä and Seppo Meri

1. Introduction

The human complement (C) system is composed of about 20 components in blood plasma (Chapter 1). Of these, many become deposited at sites of complement attack and are detected by immunohistochemical methods. C3b, C4b, and their fragments remain covalently bound on targets and components of the membrane attack complex (MAC) can become inserted into cell membranes. In addition, many components (like C1q and properdin) are stable enough and have such a high affinity for targets that they can remain in tissues for prolonged periods.

Complement deposition and activation products have been reported in affected tissues in systemic and tissue-specific autoimmune disorders. Complement deposits in glomeruli are a common finding in renal diseases, such as poststreptococcal and mesangiocapillary glomerulonephritis *(1)*. C1, C3, C9 and MAC deposits have been shown in thyroid follicular basement membranes in Hashimoto's thyroiditis and Graves' disease *(1)*. Local complement activation has been shown in the joints of rheumatic arthritis patients, in dermatomyositis lesions, and on vascular endothelium in some vasculitides. There are complement components and activation products in the intestinal lesions of patients with Crohn's disease and ulcerative colitis *(1)*. Local complement activation is also involved in urticaria, angioedema, bullous pemphigoid, pemphigus vulgaris, porphyria cutanea tarda, dermatitis herpetiformis, and psoriasis patients *(1)*.

Complement components (C1q, C3, C4, C5–C9) and activation products can be demonstrated in both myocardium and endothelium of acute myocardial lesions as a result of ischemia-reperfusion injury *(2,3)*. C3, C4, and MAC deposits have been observed in intimal thickenings and in fibrous plaques in

From: *Methods in Molecular Biology, vol. 150: Complement Methods and Protocols*
Edited by: B. P. Morgan © Humana Press Inc., Totowa, NJ

atherosclerotic lesions *(4)*, as well as in lesions of brain infarction, Alzheimer's disease, and multiple sclerosis. Complement deposition also occurs in affected tissues in several viral, parasitic, and bacterial infections.

Immunohistochemical methods based on fluorochrome- or enzyme-labeled antibodies are commonly used for the demonstration of complement components (C1q, C3, C4, C5-C9, properdin) and activation products (iC3b, C3c, C3d, or C5b-9) in tissues. The detection of complement activation products (e.g., iC3b, C3d or C5b-9) in tissues by neoepitope-specific antibodies (*see* Chapter 7) provides more convincing evidence of complement activation than the use of antibodies against native complement components. There are many soluble complement activation products and components (e.g., C1r, C1s, C3a/C3a desArg, C4a/C4a desArg, C5a/C5a desArg, Bb, D, I) which remain in the fluid phase, have a weak association, and/or are rapidly converted into inactive forms. These products are not readily detectable in tissues by immunohistochemistry.

Frozen cryostat sections of human autopsy samples or biopsies or tissues from laboratory animals can be used for immunohistochemical analysis. Some antibodies [especially monoclonal antibodies (MAbs)] do not work on paraffin-embedded tissue sections. The fixation procedure can affect the result of the immunohistochemical staining. If positive staining result cannot be obtained after testing various fixation methods (acetone, paraformaldehyde, or ethanol), one should try to use unfixed frozen sections.

Both immunofluorescence and immunoperoxidase/alkaline phosphatase-based immunohistochemical methods have been used in complement research. In general, immunofluorescence analysis is more specific, quicker to perform, but a less sensitive method than immunoenzymatic analyses. The latter is compatible with many histological staining procedures. Although immunofluorescence staining often fades within days or weeks, immunoenzymatically stained specimens remain visible for a much longer time. Some special analysis techniques (i.e., confocal microscopy) are available primarily for immunofluorescence stained specimens.

In this chapter, we will describe immunohistochemical methods generally used for the demonstration of complement activation in human and animal tissues.

2. Materials
2.1. Equipment
2.1.1. Preparation of Frozen Sections

1. Cryotome.
2. Tissue-Tek® O.C.T. tissue block embedding medium (Miles, Elkhart, IN).

3. Airtight plastic bags.
4. Dry ice or liquid nitrogen-precooled isopentane.
5. Liquid nitrogen.
6. Glass slides.

2.1.2. Immunohistochemical Stainings

1. Coplin jars.
2. Moisture chamber.
3. Cover slips.
4. Wax pencil.
5. Paper towels.
6. Pipets.

2.2. Reagents

2.2.1. Antibodies and Chemicals

2.2.1.1. SOURCES OF PRIMARY ANTIBODIES AGAINST COMPLEMENT COMPONENTS

See **Appendix** for further details.

1. Quidel, San Diego, CA.
2. Dako, Glostrup, Denmark.
3. Serotec, Oxford, UK.
4. Bio-Products Laboratories, Elstree, UK.
5. Immunotech, Marseille, France.
6. Cappel, Malvern, PA.
7. Individual investigators and research groups.

2.2.1.2. SECONDARY ANTIBODIES

Alkaline phosphatase-, peroxidase- or fluorescein isothiocyanate (FITC)-conjugated secondary antibodies (e.g., Jackson Immunoresearch Laboratories, West Grove, PA; Dako, Glostrup, Denmark).

2.2.1.3. BUFFERS AND CHEMICALS

1. Phosphate-buffered saline (PBS). Prepare 10× stock solution by dissolving 1.4 M NaCl, 21 mM KCl, 15 mM KH$_2$PO$_4$, and 82 mM Na$_2$HPO$_4$ in 1000 mL of deionized water. The pH of the final (1x) solution will be 7.4.
2. Bovine serum albumin (BSA) is Fraction V from Sigma.
3. 3-amino-9-ethyl carbazol (AEC) substrate solution (prepared fresh daily): 4 mg 3-amino-9-ethyl carbazol (AEC; Sigma) in 0.5 mL N, N-dimethylformamide. This solution is added to 9.5 mL 0.1 M sodium acetate buffer (pH 5.2) and the mixture is filtered. Add 5 µL of 30% H$_2$O$_2$ immediately before use.
4. 0.1 M sodium acetate buffer: 13.61 g sodium acetate in 800 mL distilled water. Adjust pH to 5.2 with acetic acid. Adjust final volume to 1000 mL.

5. DAB substrate solution: 6 mg 3'3-diaminobenzidine tetrahydrochloride (DAB) in 10 mL of PBS; 5 μL of 30% H_2O_2 is added just before use.

2.2.2. Fixation

2.2.2.1. PARAFORMALDEHYDE FIXATION

1. Paraformaldehyde (Sigma); 8% stock solution in deionized water. Adjusting the pH with 1 *M* NaOH to 7 and heating the solution to +60°C helps dissolve the paraformaldehyde.
2. Working solution is: 4% paraformaldehyde solution in PBS. Mix five parts paraformaldehyde (8% stock) with one part PBS (10x stock) and four parts distilled H_2O (v/v).

2.2.2.2. ACETONE FIXATION

1. Acetone chilled to –20°C by placing in a laboratory freezer.

2.2.3. Making Mounting Medium

1. Mix 2.4 g Mowiol (Calbiochem; La Jolla, CA) with 6 g of glycerol. Add 6 mL of distilled H_2O and leave for several hours at room temperature.
2. Add 12 mL of 0.2 *M* Tris (pH 8.5) and heat to 50°C for 10 min with occasional mixing.
3. Mowiol will not dissolve completely. Therefore, remove the undissolved particles by centrifugation at 5000*g* for 15 min.
4. Finally, add 1,4-diazobicyclo-[2.2.2]-octane (DABCO; Aldrich, Milwaukee, WI) to 2.5% to prevent fading of the fluorescence in the samples. Store appropriate aliquots at –20°C.
5. When thawing, allow the solution to reach room temperature before mounting. This is to prevent formation of air bubbles under the coverslip.
   Caution. There is limited evidence that Mowiol is carcinogenic in laboratory animals, therefore wear protective gloves.

3. Methods

3.1. Tissue Fixation

3.1.1. Paraformaldehyde Fixation

1. Immerse the cryostat sections in working solution for 10 min.
2. Wash the sections with PBS (x3).
   See **Note 1**.

3.1.2. Acetone Fixation

1. Immerse the cryostat sections (5 μm) in acetone (–20°C) for 5 min.
2. Rinse the sections in PBS (x3).
   See **Note 1**.

3.2. Immunofluorescence Histochemistry

3.2.1. Preparation of Frozen Tissue

3.2.1.1. LIQUID NITROGEN METHOD

1. Dissect a tissue sample (maximum size 1 cm^2 × 0.4 cm).
2. Freeze in isopentane precooled (–70°C) with liquid nitrogen for 60 s.
3. Place the sample into an airtight plastic bag and on dry ice.
4. Store at –70°C.

3.2.1.2. DRY ICE METHOD

1. Dissect a tissue sample (maximum size 0.8 cm^2 × 0.4 cm).
2. Fill a plastic mold (e.g., 1 cm × 1 cm × 1 cm) with Tissue-Tek OCT medium.
3. Place the sample in the OCT medium and put on dry ice until frozen.
4. Place the frozen block in an airtight plastic bag and store at –70°C.

3.2.2. Preparation of Cryostat Sections

1. Immobilize the frozen tissue sample (max. size 1 cm^2 × 0.4 cm) on top of the precooled cryostat chunk (e.g., –20°C) by using a few drops of Tissue-Tek.
2. Cut 5–7-µm cryostat sections on slides at –20°C (*see* **Notes 2** and **3**).
3. Allow the slides to air-dry for at least 10 min. Inadequate drying of the sections may cause them to detach from the slide during the staining procedure. Frozen sections can be stored at –20°C for weeks.

3.2.3. Immunofluorescence Staining Method for Frozen Sections

1. Immerse the cryostat sections in acetone (–20°C) for 5 min. After fixation, wash the slide with PBS (x3), remove excess liquid around the specimen. Acetone is an organic solvent and will remove a proportion of lipids and precipitate proteins.
2. Apply 100–200 µL of the appropriately diluted primary antibody. Verify that the whole section area is covered with the solution. Incubate for 20–30 min.
3. Rinse the slide with PBS (x3).
4. Remove excess liquid around the specimen.
5. Apply 100–200 µL of fluorochrome-labeled secondary antibody diluted appropriately. Incubate 20–30 min.
6. Rinse the slide with PBS (x3).
7. Remove excess liquid around the specimen.
8. Put cover slips on the slides with the help of a drop of Mowiol as the mounting medium. Avoid air bubbles.
9. Examine the slides with a fluorescence microscope.
   See **Notes 4** and **5**.

3.3. Immunoenzymatic Methods

The sample preparation and antibody incubation steps are essentially as described in **Subheading 3.2.** The primary antibody reacts with the antigen in

the tissue. Enzyme-labeled secondary antibody binds the primary antibody. The enzymes usually used are horseradish peroxidase or alkaline phosphatase. Substrate and chromogen reaction concludes the sequence and the antigen is detected by a color reaction in a tissue section in light microscopical analysis *(5)*.
     See **Note 6**.

3.3.1. Immunoperoxidase Staining Method for Frozen Sections

1. Take 3–5-µm thick frozen sections on glass slides.
2. Air dry sections.
3. Fix with acetone (–20°C) for 5 min.
4. Wash with PBS for 5 min (x3).
5. If indicated (blocking of endogenous peroxidases needed), incubate sections with 0.3% H_2O_2 in water or methanol for 5 min. In some cases, the antigen may be destroyed by H_2O_2 treatment. In these cases, the H_2O_2 treatment can only be performed after incubation with the biotinylated secondary antibody.
6. Wash with PBS for 5 min (x3).
7. Block nonspecific interactions with normal serum of the same species as the biotinylated antibody for 20 min in a moist chamber.
8. Remove excess normal serum by tapping the slides on the edge; incubate the slides with the primary antibody in a moist chamber for 30–60 min. Rinse the slides and place them in PBS for 5 min. *See* **Note 7**.
9. Incubate the slides with the biotinylated secondary antibody in a moist chamber for 30 min. Rinse the slides and place them in PBS for 5 min.
10. Incubate the slides with the avidin-biotin-enzyme complex (ABC) in a moist chamber for 30 min. Rinse the slides and place them in PBS for 5 min.
11. Develop the reaction with 3'3-diaminobenzidine tetrahydrocloride (DAB; 6 mg in 10 mL of Tris-HCl buffer (pH 7.6) and 100 µL of 3% hydrogen peroxide) for 5 min or less. (*See* **Note 8**.)
12. Stop the reaction by rinsing the slides with distilled or tap water for 10 min.
13. Counterstain with 3–5 quick dips in Mayer's hematoxylin. Wash in running tap water for 10 min.
14. Dehydrate, clear, and mount with resinous mounting medium. For AEC, delete this step and mount in aqueous medium *(5)*.
     See **Notes 9** and **10**.

3.4. Control Procedures

1. All primary and secondary antibodies should be tested in a dilution series on appropriate tissue sections to find the optimal dilution for use in future studies. *See* **Note 11**.
2. Negative controls should include the following: A negative tissue control (tissue that does not contain the relevant antigen); negative controls for primary antibodies (pre- or nonimmune antiserum and irrelevant antiserum from the same species as the primary antiserum or irrelevant monoclonal antibody of the same

isotype as the primary antibody); negative controls for secondary antibodies; buffer controls (primary or secondary antibody is replaced with the dilution buffer).

3. To control for autofluorescence, omit the immunostaining process or look at the stained sections with another filter. When a filter for FITC is used, auto-fluorescence usually appears yellow and can be seen also with filters specific for TRITC. *See* **Note 12**.

4. Positive controls should include the following: a positive tissue control (tissue that is known to contain the relevant antigen); a positive primary antibody control (antibody that is known to detect the relevant antigen).

5. The specificity of antibodies should always be checked by an independent method, e.g., by immunoblotting.

4. Notes

1. Instead of paraformaldehyde fixation, acetone, ethanol, or methanol (–20°C) can be used as fixatives. However, cellular morphology is not as well preserved as with paraformaldehyde. Acetone is an organic solvent and will remove part of lipids and precipitate the proteins.

2. Slides should be clean to improve the binding of sections. Slides can be cleaned by soaking them first in 10% HCl in ethanol (v/v) for 10 min and then in ethanol for another 10 min. Thereafter rinse the slides in H_2O twice for 10 min and immerse in ethanol for 10 min and let air-dry.

3. A useful method to make the glass surface more adherent is poly-L-lysine coating. The positively charged poly-L-lysine will bind to most surfaces and cells that in general have an overall negative charge. Prepare a solution of 1 mg/mL poly-L-lysine in distilled H_2O, immerse clean glass slides for 15 min, wash with H_2O and let air-dry.

4. Indirect immunofluorescence analysis and double-labeling technique (i.e., utilizing TRITC- and FITC-conjugated secondary antibodies) can be used for the detection of two antigens on the same frozen section. To avoid possible cross-reactions, the primary antibodies should be from different host species (i.e., from mouse and rabbit). The crossreactivity of the secondary antibodies against the primary antibodies and against the tissue antigens should be tested by control immunofluorescence stainings. It is recommended to use secondary antibodies, which are affinity-purified and preabsorbed with serum proteins of other species. Similarly, immunoenzymatic double-labeling methods (i.e., immunoperoxidase-AEC and alkaline phosphatase-nitrobluetetrazolium) can be used. In all cases the best option would be to have differentially labeled primary antibodies.

5. The indirect immunofluorescence method can be used for paraffin embedded deparaffinized tissue sections; however, paraffin-embedding will destroy many tissue antigens and especially some monoclonal primary antibodies may not work.

6. Store the stock solutions of antibodies at –70°C (–20°C) in small aliquots. Avoid repeated thawing and freezing. Certain stock solutions can be stored at +4°C for months. Sodium azide (NaN_3) is a commonly used preservative. It cannot, how-

ever, be used in the immunoperoxidase assays because it blocks the peroxidase activity.

7. The nonspecific binding of primary antibodies can be minimized by using an irrelevant coating protein. Incubate the sections, e.g., in 1–3% bovine serum albumin (BSA) in PBS or in nonfat dry milk (5–10%) in PBS for 5 min.

8. DAB is potentially carcinogenic! Wear gloves and avoid inhalation. Substitute DAB with AEC if a red end reaction is preferred.

9. Incubation times are usually 30 min (in special cases, anything from 1.5 min to 48 h). Shorter incubations are performed at +20°C or +37°C. Longer incubations are performed at +4°C in moist chambers. Prolonged incubation times may allow the use of lower antibody concentration.

10. A greasy wax pen (PAP-PEN or equivalent) can be used to draw a circle around the tissue section on slide. This procedure will minimize the amount of the antibody solution and makes the wiping around the tissue sections unnecessary.

11. The range of antibody concentrations used is usually 0.1–20 µg/mL for monoclonal antibodies. If concentrations are not known, one may start with 1/10–1/1000 dilutions for MAb cell-culture supernatants and 1/50–1/2000 dilutions for polyclonal antisera.

12. To prevent fading in the immunofluorescence staining, use antifading agents (e.g., 2.5% 1,4-diazobicyclo-[2.2.2]-octane in Mowiol; DABCO, Aldrich) in the mounting medium. In general, the immunostained samples should be analyzed and photographed as soon as possible. One should choose a quick film (400–800–1600 ASA) to avoid exposure related fading of the sample. However, quicker films have lower resolution. The utilization of high-quality digital cameras and/or confocal microscopy may reduce the fading by decreasing the exposure time. Some antihuman complement antibodies crossreact with animal complement components or activation products; e.g., MAb against human C5b-9 neoantigens, human C3c, and human properdin crossreact with corresponding porcine complement proteins (*6*). The specificity of this crossreactivity should be tested by, e.g., the immunoblotting of the purified animal complement protein.

13. Notes on interpretation and special questions related to complement stainings.

    a. The activation of the classical pathway can occur antibody-independently, i.e., C1q can bind directly to cytoskeletal intermediate filaments or mitochondrial membranes.

    b. C3ib deposition indicates an acute phase reaction of inflammation, whereas C3c deposition remains in tissues for a longer time. C3d deposition is often seen on normal human arteries and it is caused by the low-level complement activation that generally occurs.

    c. C5b-9 deposition indicates that the complement activation has occurred *in situ* in tissues, whereas SC5b-9 complexes are usually formed in circulation and thereafter deposited on cell membranes as an inactive SC5b-9 form of the MAC. However, differentiation between SC5b-9 and C5b-9 complexes cannot be made by immunohistochemistry, because no monoclonal antibody has been found to be specific for either SC5b-9 or C5b-9. These two forms of

MAC can be identified on the basis of their characteristic sedimentation behavior in sucrose density gradient analysis *(2)*.

References

1. Morgan, B. P. (1991) *Complement. Clinical aspects and relevance to disease.* 1st ed. Academic, London, UK.
2. Hugo, F., Hamdoch, T., Mathey, D., Schäfer, H., and Bhakdi, S. (1990) Quantitative measurement of SC5b-9 and C5b-9(m) in infarcted areas of human myocardium. *Clin. Exp. Immunol.* **81,** 132–136
3. Väkevä, A., Laurila, P., and Meri, S. (1993) Regulation of complement membrane attack complex formation in myocardial infarction. *Am. J. Pathol.* **143,** 65–75.
4. Niculescu, F., Rus, H. G., and Vlaicu, R. (1987) Immunohistochemical localization of C5b-9, S-protein, C3d and apolipoprotein B in human arterial tissues in atherosclerosis. *Atherosclerosis* **65,** 1–11.
5. Espinoza, C., Livingston, S., and Azar, H. (1992) Immunohistochemistry in surgical pathology: principles, methods, and quality control, in *Manual of Clinical Laboratory Immunology* ASM. pp. 277–281.
6. Jansen, H. J., Høgåsen, K., and Mollnes, T. (1993) Extensive complement activation in hereditary porcine membranoproliferative glomerulonephritis type II (porcine dense deposit disease). *Am. J. Pathol.* **143,** 1356–1365.

9

Complement Regulators and Receptors in Tissues

Juha Hakulinen and Seppo Meri

1. Introduction

Complement regulators and receptors can be identified individually in tissues by probing thin (5 μm) tissue sections fixed on a solid glass support with appropriate antibodies. The bound antibodies are visualized with a proper label that has been attached to the primary antibody itself or to a secondary antibody against the primary immunoglobulin (*see* **Note 1**). Despite the label the antibody retains its reactivity with its antigen. In immunofluorescence microscopy the label is a fluorochrome that emits visible light under exposure to exciting radiation like ultraviolet (UV) light. Washing out the extra stain reveals the sites where antibody has contacted the antigen. These sites appear as brightly illuminated areas in contrast to the dark background (*see* **Notes 2** and **3**). The color of the light or fluorescence depends on the chromogen used: Fluorescein iso-thiocyanate (FITC) excites at 495 nm and emits green light (peak at 525 nm) whereas tetra-methyl rhodamine-iso-thiocyanate (TRITC) absorbs maximally at 552 nm and gives red fluorescence (peak at 570 nm). To detect fluorochrome labels special microscopes equipped with an UV-light source and appropriate filters are required. In immunoenzymatic stainings enzymes conjugated to antibodies react with appropriate substrates and generate color deposits on the samples. Distribution of the target antigens can thus be examined with a light microscope (*see* **Note 4**).

Several complement inhibitor molecules have been described in human cells and tissues. It seems that every normal cell in the body expresses at least one of the complement regulators on their surface. These regulators include complement receptor type 1 (CR1; CD35), membrane cofactor protein (MCP; CD46), decay accelerating factor (DAF; CD55) and protectin (CD59) *(1–3)*. CR1, MCP and DAF inhibit complement at the level of C3.

From: *Methods in Molecular Biology, vol. 150: Complement Methods and Protocols*
Edited by: B. P. Morgan © Humana Press Inc., Totowa, NJ

CD59 is an 18–24 kDa glycoprotein that can directly inhibit the formation of MAC (*see* **Note 5**). Both DAF and CD59 are anchored to cell membranes via a glycophosphoinositol (GPI) moiety. Soluble plasma proteins vitronectin (S-protein) and clusterin (apo J, sp40,40) bind to terminal complement complexes (TCC) inhibiting their binding to plasma membranes. However, at least vitronectin can often be found sometimes on cell surfaces bound to TCC complexes *(4)* (*see* **Note 6**). Complement receptors CR1 (CD35), CR2 (CD21), CR3 (CD11b/CD18), and CR4 (CD11c/CD18) bind to activation fragments of complement and have responsibilities, e.g., in the processing and clearance of complexes containing C3 and C4 *(5–8)*.

2. Materials

1. Liquid nitrogen.
2. Tissue-Tek O.C.T (Miles Inc., Elkhart, IN).
3. 2-methylbutane (isopentane).
4. Paraformaldehyde (Sigma; St. Louis, MO). Prepare 8% stock solution in deionized water. Adjusting the pH with 1 M NaOH to 7 and heating of the solution to +60°C helps dissolving of the paraformaldehyde. **Caution:** paraformaldehyde is toxic by inhalation and by skin contact. It is carcinogenic and may cause DNA damage.
5. Cold (–20°C) acetone for fixation.
6. Phosphate-buffered saline (PBS). Prepare a stock solution (10×). Dissolve 80 g NaCl, 2 g KCl, 2 g KH_2PO_4, and 14.6 g $Na_2HPO_4 \times 2H_2O$ in 1000 mL of deionized water. Do not adjust the pH, the final (1×) solution will be pH 7.4.
7. Bovine serum albumin (BSA) Fraction V (Sigma).
8. Mounting medium: Mix 2.4 g Mowiol 4-88 (Calbiochem; La Jolla, CA) or Gelvatol 20-30 (Gelvatol 20-30 or an equivalent product Airvol 205 can be obtained from Air Products Nederland B.V. Utrecht) with 6 g of glycerol. Add 6 mL of dH_2O and 12 mL of 0.2 M Tris, pH 8.5, stir overnight to mix (*see* **Note 7**). Heat to 50°C for 10 min. Because Mowiol will not dissolve completely remove the undissolved particles by centrifugation at 5000g for 15 min. Finally, add 1,4-diazabicyclo-[2.2.2]-octane (Dabco; Aldrich, Milwaukee, WI) to 2.5% to prevent fading of fluorescence in the samples. Store appropriate aliquots at –20°C. The mounting medium will last for 2 wk at room temperature. **Caution:** There is limited evidence that Mowiol is tumorigenic in laboratory animals so wear protective gloves.
9. Film: Black and white film or film for color slides, ASA 400.

3. Methods
3.1. Preparing Frozen Tissue Sections

There are several methods for preparing tissue samples for cell staining. Most commonly the specimens are fixed in formalin and embedded in paraffin. However, preparation of the tissue for sectioning by freezing the specimen with the help of liquid nitrogen is perhaps the gentlest method that preserves the epitopes for complement regulators in the sample.

1. Preparing small pieces (<1 mm) for sectioning: Place the pieces in physiological buffer and allow to sediment into the tip of an Eppendorf tube. Remove the buffer as completely as possible. Ice crystals will make the sectioning difficult.
2. Add a drop (ca 200 μL) of the mounting medium Tissue Tek O.C.T or equivalent on the sample and dip the tube into liquid nitrogen until the Tissue Tek solidifies. Turn the tube upside down and snap it firmly against a lab bench to remove the frozen cone. The specimens *must not melt after freezing* and must be immediately placed in an appropriate container on dry ice or in liquid nitrogen (*see* **Note 8**). The samples can be stored at –70°C or immediately sectioned in a refrigerated microtome (cryostat). Freezing of larger specimens directly in liquid nitrogen may destroy some of the tissue architecture. Therefore, larger specimens are preferably frozen by soaking them in isopentane cooled with liquid nitrogen (*see* **Note 9**). Immerse the sample in isopentane until it stops bubbling.
3. The sectioning requires a microtome that can be cooled to –20°C. In general, thinner sections are better for staining, mostly sections from 5 to 10 μm are used for immunofluorescence staining. Some tissues are especially difficult to fix on a glass slide. Usually, frozen cryostat sections stick to a glass plate if only very lightly touched with the slide. However, depending on the tissue sample some sections have a tendency to come off the glass support during the staining. In these cases the slides can be specifically treated to make the glass surface more adherent.

3.2. Treatment of the Slides

3.2.1. Cleaning of Slides

It is essential that the slides are clean to improve the binding of sections. The glasses can be washed in acid alcohol solution.

1. Soak the slides in 10% HCl ethanol (v/v) for 10 min.
2. Transfer the slides to ethanol for another 10 min.
3. Keep the slides in H_2O 2 × 10 min.
4. Finally, immerse the slides in ethanol for 10 min and let them air-dry.

3.2.2. Poly-lysine Coating

Another method to make the glass surface more adherent is to coat it with poly-L-lysine. Lysine is an amino acid with a positively charged side chain. Poly-L-lysine will bind to most surfaces and cells that have an overall negative charge.

1. Prepare a solution of poly-L-lysine, to 1 mg/mL, in distilled H_2O.
2. Immerse clean glass slides in poly-L-lysine for 15 min.
3. Wash the coated slides with dH_2O.

4. After drying the poly-L-lysine coated slides will be ready for use to mount the cryostat sections. Store the slides at room temperature.

3.3. Fixation of Cryostat Sections

Fixation of the tissue samples should preserve the tissue architecture and antigen distribution as native as possible.

3.3.1. Paraformaldehyde Fixation

Paraformaldehyde will crosslink the proteins in the sample preserving the cell membrane integrity.

1. Prepare 4% paraformaldehyde solution in PBS by mixing five parts paraformaldehyde (8%), 1 part PBS (10×) and four parts dH$_2$O (v/v).
2. Immerse the cryostat sections in the fixative for 10 min.
3. Rinse the sections with PBS.

3.3.2. Acetone Fixation

Acetone is an organic solvent that will remove lipids and precipitate the proteins exposing also the intracellular antigens.

1. Incubate the cryostat sections in neat acetone at –20°C for 1–10 min depending on the nature of the antigen (*see* **Note 10**).
2. Rinse the sections with PBS.

3.4. Staining for Immunofluorescence Microscopy

1. Ensure that there are enough sections for the assay planned, including all the controls. All samples within an assay should be processed equally (*see* **Note 13**).
2. Fix the tissue sections with paraformaldehyde or acetone as described above (*see* **Note 14**).
3. Rinse the fixed sections with PBS three times.
4. Block the nonspecific binding of antibodies in the paraformaldehyde or acetone fixed sections with an irrelevant protein by incubating the sections in 3–5% BSA in PBS for 5 min.
5. Prepare an appropriate dilution of the primary antibody in the BSA/PBS (*see* **Note 15**).
6. Wipe the additional blocking buffer around the sections with a tissue paper. Add the primary antibody on the section and incubate at RT for 30 min in a humidified chamber (*see* **Note 16**). The sections should not be let to dry out.
7. Wash the samples three times for 5 min in PBS.
8. Prepare a dilution (1/50–1/200) of the FITC-labeled antibody in BSA/PBS. Add 50 µL on the sections and incubate for another 30 min at RT.
9. Repeat the washing.

10. Finally, drain the extra buffer off and mount the sections by dropping 10-20 µL of mounting medium (Mowiol/DABCO) on the section and placing the cover slip on the sample (*see* **Note 17**).

4. Notes

1. Indirect immunofluorescence has advantages over direct IF. Usually, the fluorescence is brighter because several fluorochrome-containing secondary antibodies are bound to a single primary antibody. Also, it is possible to detect with one labeled antiimmunoglobulin conjugate several different antigens. In general, the antibodies or sera are diluted into a physiological phosphate-, TRIS-, or veronal-buffer. The dilution of the antiserum or concentration of the antibody depends, e.g., on the affinity of the reagent. High-affinity antibodies can be diluted more than low-affinity antibodies. For monoclonal antibodies the concentration usually ranges from 0.2 to 20 µg/mL.
2. GPI-anchored complement inhibitors (CD59 and CD55) have a tendency to become clustered on cell membranes instead of being smoothly and evenly distributed. Thus, they usually appear as small dots on cell membranes and on cell extensions.
3. CD59 and CD55 staining may decrease in areas of tissue necrosis and on apoptotic cells.
4. The advantage of fluorescent stains is their high resolution. It is also possible to stain two different antigens in the same sample by using double labeling with reagents that emit different colors of light (FITC and TRITC, for example). Fading is a serious disadvantage of immunofluorescence. However, this can be adequately controlled with antifading agents. Some think P-phenylenediamine (PPD; Sigma) is the best antifading agent available at a concentration of 0.1%. However, PPD is considered carcinogenic and should be handled with care. Diazabicyclo-octane (DABCO) has lower toxicity but is somewhat less effective that PPD. Vectashield is an antifade commercially available from Vector Co. (Burlingame, CA). Enzyme labels are more sensitive and can be examined with a light microscope. Also, the staining is permanent. Some cells may demonstrate endogenous enzyme activity.
5. In areas of C activation in tissues the epitopes of C regulators for the MAb used may become obscured. Thus, binding of CD59 to membrane attack complex (MAC) complexes may cover CD59 epitopes for some antibodies.
6. Soluble regulators of C (C4bp, clusterin, and vitronectin) can be visualized in tissues if they have become strongly bound. On the other hand, e.g., factor H and C1INH seldomly remain firmly bound in tissues.
7. The fluorescent emission is greater at alkaline than acid pH. Occasionally, the molecules in the sample will also be excited to some extent by UV-radiation and emit visible light. This so-called autofluorescence should not be confused with true fluorescence.
8. For small blocks of samples 2 mL vials are adequate. The vials can be cooled by simply dipping them into liquid nitrogen. Remember to mark the vials properly describing the contents as thoroughly as possible.

9. To cool the isopentane in liquid nitrogen two containers are needed, a larger insulated vessel for liquid nitrogen and a smaller one for isopentane. Put the smaller container into the larger one and pour some liquid nitrogen into the outer vessel. Pour some isopentane into the smaller container. It is wise to keep some isopentane precooled in a freezer. Isopentane will solidify eventually, however, it can be easily melted by inserting a metallic bolt or a bar into it.

10. Staining for GPI-anchored complement inhibitors (CD55 and CD59) has special features that need to be recognized. Organic solvents, like acetone, may remove a proportion of these relatively loosely membrane associated phospholipid-tailed proteins. For best performance, use short incubation times (1–3 min) in acetone, paraformaldehyde or stain the sections unfixed. Formalin-fixation and embedding in paraffin will destroy epitopes for many C regulator antibodies.

11. Sometimes CD55 and CD59 staining appears to occur on certain connective tissue or intracellular filaments. These observations should be interpreted with caution because they can be due to crossreactivities of the antibodies. For example, the YTH53.1 rat MAb against human CD59 binds to intermediate filaments on certain types on epithelial cells.

12. White blood cells and placental cells contain receptors (Fc-gamma-RI, -RII, and -RIII) that bind immunoglobulin Fc-part specifically. The receptors can be blocked by adding 1% heat inactivated normal serum (35 min at 56°C) to the blocking and diluting buffers.

13. A convenient way is to draw a circle around the sections before the staining assay with a greasy marking pen (PAP-PEN or equivalent) to minimize the amount of precious antibody required. PAP-PEN (Daido Sangyo Co. Ltd, Japan, or Sigma, St. Louis, MO) makes a water-repellant circle keeping liquid pooled in a droplet. It also makes wiping around the sections unnecessary .

14. The choice of a fixative and the time of fixation as well as the temperature are usually determined empirically. Some agents may destroy the reactive epitope. If paraformaldehyde is used, permeabilization of the cells is required if the antigen is intracellular. For the immunofluorescence staining of DAF, CD59 and MCP the samples can be used native or fixed either with 4% paraformaldehyde or acetone at –20°C.

15. The dilution of the antiserum or the concentration of a monoclonal primary antibody depends on factors like antibody affinity and should be tested individually. In general, the concentrations for monoclonal antibodies range from 0.2 to 20 µg/mL. The amount of the antibody dilution required depends on the area of the specimen, normally 50 µL is sufficient.

16. Sometimes prolonging the incubation time overnight at +4°C may increase the sensitivity.

17. When thawing let the Mowiol/DABCO solution reach room temperature before mounting to prevent the formation of air bubbles under the coverslip. Sometimes the salts in the buffer may precipitate forming crystals in the sample this can be prevented by simply dipping the samples briefly in distilled water just before use.

18. To rule out nonspecific binding of antibodies the reactivity of all components in the assay should be tested individually with the specimen. The binding of the

primary antibody should always be compared to another antibody of same subclass that has no antigenic reactivity with the specimen. If antiserum is used a preimmunization serum from the same animal or at least from the same species should be tested. A positive control that is known to react with the primary antibody should also be included. Usually, in antiserum the concentration of nonspecific antibodies is low and crossreactivity can be diluted out. Monoclonal crossreactive antibodies are trickier because the crossreactive paratope cannot be usually diluted out (*see* **Note 11**). Nonspecific binding can be reduced by letting the sample to react with an irrelevant protein like bovine serum albumin in the buffer. Albumin will reduce binding to the nonspecific binding sites. Furthermore, some cells and tissues may contain receptors for immunoglobulins that may bind the antibodies specifically through interactions that do not involve the antigen binding site but the Fc-part of the antibody (*see* **Note 12**).

19. Immunohistochemistry is a useful and cost-effective way of obtaining informative results. However, be aware of the sometimes very high publication charges of color prints.

References

1. Seya, T., Turner, J. R., and Atkinson, J. P. (1986) Purification and characterization of a membrane protein (gp45-70) that is a cofactor for cleavage of C3b and C4b. *J. Exp. Med.* **163,** 837–855.
2. Nicholson-Weller, A., Burge, J., Fearon, D. T., Weller, P. F., and Austen, K. F. (1982) Isolation of a human erythrocyte membrane glycoprotein with decay-accelerating activity for C3 convertases of the complement system. *J. Immunol.* **129,** 184–189.
3. Sugita, Y., Nakano, Y., and Tomita, M. (1988) Isolation from human erythrocytes of a new membrane protein which inhibits the formation of complement transmembrane channels. *J. Biochem.* (Tokyo) **104,** 633–637.
4. Høgåsen, K., Mollnes, T. E., and Harboe, M. (1992) Heparin-binding properties of vitronectin are linked to complex formation as illustrated by in vitro polymerization and binding to the terminal complement complex. *J. Biol. Chem.* **267,** 23,076–23,082.
5. Lowell, C. A., Klickstein, L. B., Carter, R. H., Mitchell, J. A., Fearon, D. T., and Ahearn, J. M. (1989) Mapping of the Epstein-Barr virus and C3dg binding sites to a common domain on complement receptor type 2. *J. Exp. Med.* **170,** 1931–1946.
6. Morgan, B. P. (1995) Complement regulatory molecules: application to therapy and transplantation. *Immunol. Today* **16,** 257–259.
7. Krych, M., Atkinson, J. P., and Holers, V. M. (1992) Complement receptors. *Curr. Opin. Immunol.* **4,** 8–13.
8. Meri, S. and Jarva, H. (1998) Complement regulation. *Vox. Sang.* **74,** 291–302.

10

Measurement of C3 Fragment Deposition on Cells

O. Brad Spiller

1. Introduction

Measurement of complement-mediated cytolysis, described in Chapter 5, is the index most frequently used for assessing aspects of complement activation or regulation. However, other consequences of complement activation are probably of more relevance in vivo. Complement plays a major role in enhancing the generation of specific antibodies and provides an interface between the humoral and cellular immune responses (1–4). Complement also increases the binding of phagocytes to target cells by opsonizing the complement activating surface and generates chemoattractants and anaphylatoxins to recruit nearby phagocytes (5,6). Through these mechanisms, complement not only provides a first line defence against invading pathogens but also acts as the glue that holds the major components of the immune system together. These opsonic, chemoattractant, and anaphylactic activities are largely properties of C3 (and C5) activation fragments. Furthermore, large amounts of C3 fragments can be deposited on the surface of cells without causing significant amounts of lysis. Measurement of C3-deposition is therefore a more sensitive and probably more relevant indicator of complement activation than measurement of cytolysis.

Numerous C3 regulators have been identified. These include decay-accelerating factor (DAF), membrane cofactor protein (MCP), complement receptor 1 (CR1), factor H (fH), rodent complement-regulatory region Y (Crry), and factor I (fI) (7). With the exception of fI, all of these can be localized to the regulators of complement activation (RCA) gene cluster in each species. The large number of C3 regulators underscores the importance of C3 regulation. The methods described here are designed to assess altered expression or effectiveness of C3 regulators, as well as increases or decreases in complement activation following incubation of cells with viruses, cytokines, or other substances.

From: *Methods in Molecular Biology, vol. 150: Complement Methods and Protocols*
Edited by: B. P. Morgan © Humana Press Inc., Totowa, NJ

C3 contains a highly reactive internal thiolester bond that becomes exposed when C3a is cleaved from the α-chain of C3 during activation, conferring on C3b the capacity to form covalent bonds with any free hydroxyl or amine groups in close proximity. As a result, C3b is covalently attached to the surface of the cell or particle that triggered the activation. Regardless of which activation pathway triggers the initial cleavage of C3, cell-surface bound C3b can then form a C3 convertase with factor B and properdin, which generates more cell surface C3b (an amplification loop), as well as cleaving C5 to initiate the common terminal pathway (Chapter 1).

There are two basic methods for controling complement activation at the level of C3: accelerate the decay of C3-convertases or cleave C3b to inactive fragments. Decay-accelerating activity does not alter C3b, but dissociates C3 convertases and inhibits reformation. fI cleaves the α-chain of C3 at multiple sites rendering it inactive. fI cleavage of C3 is inefficient, and requires a C3-binding cofactor for efficiency (MCP, Crry, CR1, or fH). Following the first cleavage of C3b to iC3b, MCP and fH cease to bind to iC3b and do not catalyze further cleavage. The relative molecular mass of C3b and iC3b under nonreducing conditions is almost identical because the released fragment (C3f) has a molecular mass of only 2 kDa and the remaining portions of iC3b remain tethered by disulphide bonds. However, under reducing conditions, the 103 kDa α'-chain of C3b dissociates into the 63 kDa and 40 kDa fragments of iC3b. The β-chain of C3 is not cleaved by fI and has an apparent molecular mass of 75 kDa under reducing and nonreducing conditions. The cofactor activity of CR1 is different from the other cofactors in that it catalyzes a second cleavage of C3 α'-chain by fI. The second cleavage releases a 140-kDa fragment (under nonreducing conditions), referred to as C3c, consisting of the β-chain and the portions of α-chains that are attached to the β-chain by the cysteine bonds. The remaining portion of the C3 α-chain covalently bound to the cell is a 43-kDa fragment, referred to as C3dg.

The best reagent for assessing C3 deposition is one that will not recognize native C3, but has a high affinity for C3b and/or iC3b. Many monoclonal antibodies specific for C3 fragments are available commercially, and it is important to select antibodies appropriate to the fragment of interest *(8–10)*. The cleavage of C3b to iC3b releases only the 2-kDa C3f peptide, and as a result the majority of anti-C3b antibodies also bind iC3b. Among these, most detect the C3c fragment and often these are listed as anti-C3c antibodies. A minority of anti-C3b or anti-iC3b antibodies will also recognize C3dg *(9–11)*. This means that it is possible to use anti-C3c and anti-C3dg antibodies in combination to measure the cofactor activity of soluble CR1 for the third cleavage of iC3b.

The experimental design considerations that are important for measurement of C3 fragment deposition are very similar to those outlined in Chapter 5 for complement-mediated lysis. As with cytolysis assays, setting a baseline level of C3-deposition on the control cells is important. However, unlike cytolysis assays, the value of this baseline is arbitrary. The baseline should be chosen based upon whether the investigator anticipates an increase or a decrease in C3 deposition. In general, the basic concerns of the investigator include the following.

1. Choosing the right method of complement activation and source of serum to be used.
2. Achieving an adequate baseline level of activation. There are many different C3 regulators on most cell types and it may be necessary to specifically inhibit one or more to achieve a good baseline activation.
3. Including adequate controls. Flow cytometry is used to measure the C3 deposition, therefore it is possible that the blocking and/or sensitizing antibodies may increase the background fluorescence of the cells by crossreacting with the secondary antibody.
4. Ensuring that enough replicates are performed to allow statistical analysis.

Sensitizing or blocking antibodies used to increase complement activation may be detected by the secondary antibody used to detect anti-C3 binding, causing an artefactual high background. Therefore, the experimental design should include appropriate controls to measure the background binding of naturally occurring antibodies in the serum or crossreaction of any sensitizing or blocking antibodies with the secondary antibodies used. This can be achieved by using sensitizing/blocking antibodies from a species distinct from that in which the anti-C3 is raised and detecting with species-specific secondary antibodies. If it is impossible to avoid using a sensitizing or blocking antibody from the same species as the anti-C3 antibody, then it will be necessary to directly conjugate the anti-C3 antibody to biotin or FITC to avoid an unacceptable background binding of the secondary antibody. In the assay, the background for each sample is assessed by measuring the fluorescence after incubating the complement-attacked cells with a nonspecific isotype matched control.

2. Materials

1. Cell medium appropriate for propagation of the cells used (*see* **Note 1**).
2. 24-well plates (for adherent cells) or round-bottom 96-well plates (for nonadherent cells). [Purchased from Gibco BRL, Paisley, UK and ICN Chemical Ltd., Oxford, UK, respectively.]
3. A monoclonal or polyclonal antibody reactive with the relevant fragment of C3 from the species of interest. This antibody is referred to from this point on as the primary antibody.
4. A species and isotype-matched control antibody that does not react with C3 or the test cells.
5. A phycoerythrin-conjugated secondary antibody which reacts specifically with the primary antibody.

6. Veronal-buffered saline (VBS) to dilute the serum used (250 µL per well). This should be made fresh each day. Sold as Complement fixation diluent tablets (1 tablet makes 100 mL) by Oxoid (Basingstoke, UK). Gelatin (0.1%) should be added to the VBS to minimize nonspecific cytolysis (termed GVB^{2+}).

7. Fresh serum as a source of complement. To avoid cell loss through complement-mediated cytolysis during complement attack, it is possible to use serum depleted for one of the terminal complement components by affinity chromatography (Chapter 4). This depletion will not influence C3 deposition but will preserve cells intact for analysis.

8. Tubes suitable for flow cytometry.

9. Flow cytometry (FC) solution: phosphate-buffered saline (PBS) containing 1% bovine serum albumin (BSA) and 15 mM ethylenediaminetetraacetic acid (EDTA).

10. Optional: complement regulator blocking monoclonal antibodies (these are routinely used at 10 µg/mL, and usually with 200 µL of the diluted antibody per well) and/or sensitizing antibody (optimal amount will vary from cell type to cell type).

3. Methods

3.1. C3 Deposition on Adherent Cells

1. Seed cells into a 24-well tissue-culture plate in such a manner that they will be 70–90% confluent 24–48 h later (*see* **Notes 2–4**). Volume of cell medium should be no less than 0.5 mL per well. It is recommended that each condition to be tested be set up in three identical wells so that statistical analysis can be performed. If a time course is to be analyzed, then it is recommended that all conditions be set up to mature at the same time, rather than testing complement challenge at multiple intervals. If the number of cells are significantly different between the wells, then the C3 deposition will also be artefactually affected. It is important to have an additional well for each condition that may be disaggregated and counted just prior to complement attack.

2. Aspirate the medium from the cells and incubate the cells with 0.2 mL of the optimized concentration of sensitizing and/or blocking antibodies diluted in serum-free cell medium in the CO$_2$ incubator for 15 min. *See* **Note 5**.

3. Aspirate the medium from the cells. Wash once with 0.25 mL serum-free DMEM per well. Then add 0.25 mL of the serum diluted to the appropriate concentration in VBS. Incubate at 37°C for 10–30 min (*see* **Note 6**).

4. Check under the microscope to ensure that the cells have not been released from the plate during the incubation with complement. Aspirate the diluted serum from the cells and rinse off unbound proteins with three washes in cold PBS. Aspirate all of the PBS from the cells and add 225 µL of cold FC solution. Allow the cells to incubate with the FC solution until they detach from the plate. The release process may be accelerated by gently flushing with a 1-mL pipet. Following release, cells are harvested and kept on ice.

5. In the case of a large number of samples, incubate the primary antibody with the cells in a 96-well plate. For smaller sample numbers, the antibody incubations can be performed in the tubes used for FC. The harvested cells are split into two equal portions and incubated with anti-C3 antibody or the isotype-matched control antibody for 25 min on ice. Use primary anti-C3 monoclonal antibodies at a concentration of 5 µg/mL and polyclonal antiserum at a dilution of 1:100–1:500.

6. Spin down the cells by centrifugation at 800–1000*g* for 3 min. Wash the unbound primary antibody from the cells by three successive washes in 200 µL of ice-cold FC solution. Resuspend the cells in 100 µL of appropriate phycoerythrin-labeled secondary antibody diluted in FC solution and incubate on ice for 25 min.

7. Wash the unbound secondary from the cells by washing as in **step 6**. Resuspend the cells in 300 µL of FC solution and analyze by FC.

8. FC analysis: Set the forward and side-scatter settings to ensure efficient gating of intact cells and exclusion of debris (use log not linear scales). Set the baseline fluorescence (FL2 when using phycoerythrin) using cells not exposed to complement and incubated with the nonspecific isotype matched control primary antibody. Background fluorescence is set within the first log. Measure the fluorescence of cells exposed to maximum complement activation and incubated with the anti-C3 primary antibody. Ensure that the fluorescence of this sample is not distorted against the far right of the histogram as this will decrease the sensitivity of the assay.

9. For each sample, incubations with specific anti-C3 antibody (A) and nonspecific control antibody (B) are measured. The specific C3-deposition for each sample is calculated as (A–B). This background subtraction is performed for each sample and the mean and standard deviations for triplicate measurements in each separate set of conditions is calculated after background subtraction.

3.2. C3-Deposition Measurement for Nonadherent Cells

1. Pellet the cells by centrifugation (3 min at 1000*g*), discard the supernatant, and resuspend the cells in 0.2 mL of the optimized concentration of sensitizing and/or blocking antibodies diluted in serum-free cell medium. Incubate at 37°C for 15 min. *See* **Note 5**.

2. Pellet the cells by centrifugation (3 min at 1000*g*), discard the supernatant, and wash once by resuspending the cells in 0.25 mL of serum-free cell medium. After washing, resuspend the cells in 0.25 mL of serum diluted to the appropriate concentration in GVB^{2+}. Incubate at 37°C for 1 h. *See* **Note 6**.

3. Pellet the cells by centrifugation (3 min at 1000*g*), discard the supernatant and resuspend the cells in 200 µL of FC solution. Repeat this wash step two more times and then resuspend the cells in 225 µL FC solution. Continue from **step 5** in procedure **3.1**.

4. Notes

1. Make sure that any fetal calf serum or equivalent additive used to propagate the cells has been heat inactivated at 56°C for 1 h prior to use.

2. If the experimental design involves virus infection of the cells or incubation with cytokines or other cell activators, take into consideration the length of time required prior to complement challenge and the effects of the conditions on cell propagation.
3. The recovery period is important only if the method of subculture involves the use of enzymes to subculture the cells. If nonenzymatic methods are used for subculturing (i.e., EDTA) then it is only necessary to wait until the cells are firmly attached to the plate prior to complement attack.
4. Experiments that are designed to test the protective nature of transfected cDNA that are under selective maintenance to make stable transfectants should have appropriate controls. Some selective reagents, such as Hygromycin B, have been shown to enhance the complement activating ability of cells even if removed a few hours before. An appropriate control for this would be comparison to the same cell line transfected in the same manner for a known noncomplement activator or regulator (i.e., HLA class I).
5. Make sure that the sensitizing or blocking antibodies are not derived from the same species as the anti-C3 antibody used, as flow cytometry analysis will be impossible. If this cannot be avoided, then use a fluorescently labeled or biotinylated anti-C3 antibody for FC.
6. C3 deposition is part of an amplification pathway and the length of time used for the complement attack depends on the rate of C3 deposition. In some systems, sufficient C3 is deposited within the first 10 min, whereas in other systems, up to an hour may be required for sufficient C3 deposition. This variable is best determined by preliminary experiments.

References

1. Carroll, M. C. (1998) The role of complement and complement receptors in induction and regulation of immunity. *Annu. Rev. Immunol.* **16,** 545–568.
2. Fearon, D. T. and Carter, R. H., (1995) The CD19/CR2/TAPA-1 complex of, B., lymphocytes: linking natural to acquired immunity. *Annu. Rev. Immunol.* **13,** 127–149.
3. Thornton, B. P., Vetvicka, V., and Ross, G. D. (1996) Function of C3 in a humoral response: iC3b/C3dg bound to an immune complex generated with natural antibody and a primary antigen promotes antigen uptake and the expression of co-stimulatory molecules by all B cells, but only stimulates immunoglobulin synthesis by antigen-specific B cells. *Clin. Exp. Immunol.* **104,** 531–537.
4. Brahmi, Z., Csipo, I., Bochan, M. R., Su, B., Montel, A. H., and Morse, P. A., Jr. (1995) Synergistic inhibition of human cell-mediated cytotoxicity by complement component antisera indicates that target cell lysis may result from an enzymatic cascade involving granzymes and perforin. *Nat. Immun.* **14,** 271–285.
5. Aderem, A. and Underhill, D. M. (1999) Mechanisms of phagocytosis in macrophages. *Annu. Rev. Immunol.* **17,** 593–623.
6. Gerard, C. and Gerard, N. P. (1994) C5A anaphylatoxin and its seven transmembrane-segment receptor. *Annu. Rev. Immunol.* **12,** 775–808.

7. Morgan, B. P. and Harris C. L. (1999) *Complement Regulatory Proteins*. Academic, London, UK, pp. 41–136 and pp. 226–242.

8. Kemp, P. A., Spragg, J. H., Brown, J. C., Morgan, B. P., Gunn, C. A., and Taylor, P. W. (1992) Immunohistochemical determination of complement activation in joint tissues of patients with rheumatoid arthritis and osteoarthritis using neoantigen-specific monoclonal antibodies. *J. Clin. Lab. Immunol.* **37,** 147–162.

9. Mollnes, T. E. and Lachmann, P. J. (1987) Activation of the third component of complement (C3) detected by a monoclonal anti-C3'g' neoantigen antibody in a one-step enzyme immunoassay. *J. Immunol. Meth.* **101,** 201–207.

10. Aguado, M. T., Lambris, J. D., Tsokos, G. C., Burger, R., Bitter-Suermann, D., Tamerius, J. D., et al. (1985) Monoclonal antibodies against complement 3 neoantigens for detection of immune complexes and complement activation. Relationship between immune complex levels, state of complex levels, state of C3, and numbers of receptors for C3b. *J. Clin. Invest.* **76,** 1418–1426.

11. Iida, K., Mitomo, K., Fujita, T., and Tamura, N. (1987) Characterization of three monoclonal antibodies against C3, with selective specificities. *Immunology* **62,** 413–417.

11

Screening for Complement Deficiency

Ann Orren

1. Introduction

Complement (C) deficiencies are not common and laboratory screening for clinical reasons generally involves examining relatively few samples. Nevertheless, because of the range of tests, complete testing may mean carrying out many different assays. In contrast, the determination of the prevalence of genetic complement deficiencies in populations requires a very few specific methods that are suitable for dealing with a large number of samples.

1. Who should be screened? A guide to selection of subjects for screening for clinical reasons is discussed below. In fact, it is the patient's history, obtained before any blood is taken, that is often the correct indicator of the type of complement deficiency that is present. Thus consideration of the clinical indications for testing for deficiency forms part of the method of testing for deficiency.
2. What deficiencies will be discussed? The C system comprises the complement proteins, the regulators of complement activity, and the complement receptors. The regulators and receptors play key roles in complement function, and deficiencies of regulators lead to abnormal function and frequently secondary deficiency of certain complement proteins. In this chapter, I will discuss screening for genetic deficiencies of those serum proteins whose absence in the circulation leads to absent, or markedly reduced, function of one of the complement pathways, that is those deficiencies detected as primary or secondary deficiencies of one or more complement proteins.
3. What screening methods? If the clinical history, and the patient's circumstances, strongly point in a particular direction, it may be appropriate to test initially for particular component(s). Nevertheless, the first screening methods for complement deficiency should normally be those functional methods that investigate total complement function, rather than methods directed at identifying individual

From: *Methods in Molecular Biology, vol. 150: Complement Methods and Protocols*
Edited by: B. P. Morgan © Humana Press Inc., Totowa, NJ

deficiencies. Results of the classical pathway (CP) and alternative pathway (AP) tests, together with the subject's history are used to decide what further tests, if any, are required to make a diagnosis of whether a particular component or regulator is deficient. Many methods for further testing are detailed elsewhere in this book, and will be cross-referenced. As some methods required to make the complete diagnosis are very specialized, it is frequently better to refer to a laboratory that uses particular methods than to try to set up all methods in one laboratory.

Laboratory methods appropriate to initial screening will be given there. These include hemolytic agarose plates for screening for CP and AP activity, respectively (Subheadings 3.5. and 3.6.); a special adaptation of the hemolytic plates for simultaneously testing for several components (Subheading 3.7.); comments and references to special methods required for individual deficiencies such as properdin deficiency and a brief discussion of analyzing DNA to determine specific molecular defects.

Complement deficiencies may be genetic or acquired. The importance of acquired deficiencies is that they are frequently reversible. Only chronic serious acquired deficiencies tend to be associated with the long-term increased susceptibility to infections seen in the genetic deficiencies. Some deficiencies of complement function are secondary to a genetic deficiency of a complement regulator protein such as in C1 esterase inhibitor deficiency (Chapter 12), or Factors I (fI) or H (fH) deficiency. Complement deficiencies often lead to characteristic disease patterns and patients' histories frequently indicate which complement pathways may be affected (**Table 1**).

1. Immune complex and autoimmune disease. The classical complement system protects against the formation of immune complexes so that deficiencies of classical components and C3 are associated with immune complex (IC) diseases, such as systemic lupus erythematosus (SLE) *(1,2)*. The presentation in deficient patients is somewhat different form the general picture. The female preponderance is less, there is a younger age at presentation, there is less renal disease, and the patients are susceptible to infection with encapsulated bacteria *(3)*. The susceptibility to IC disease is strongest with C1q deficiency, less with C4 deficiency (C4D), and least with C2 deficiency (C2D), and some C4D and C2D patients are completely healthy. Patients with C3 deficiency (C3D) also frequently suffer IC complex diseases, but problems with encapsulated bacterial infections frequently tend to dominate.
2. Increased susceptibility to systemic infections with encapsulated bacteria. This occurs in all early component deficient patients, whether or not they suffer from IC diseases, and the most common organisms are Streptococcus pneumonia, Hemophilus influenzae, or Neisseria meningitidis *(3)*.
3. Angioedema is caused by C1 inhibitor deficiency (Chapter 12).
4. Increased susceptibility to repeated pyogenic infections. C3 deficiency syndromes, as well as deficiencies of the CP components can present in this way *(3)*.

C3 deficiency syndromes can arise because of primary C3D *(4)*, or secondary to fI or H deficiencies *(5,6)*, or the presence of C3 nephritic factor *(7)*. The patients have not only markedly reduced lytic complement activity, but also poor opsonophagocytic function, and are susceptible to recurrent and often severe gram-positive pyogenic infections, as well as systemic gram-negative infections, including Neisserial infections.

5. Glomerulonephritis, membrano-proliferative glomerulonephritis, or mesangio-capillary glomerulonephritis are found in patients with C3 deficiency syndromes particularly caused by C3 nephritic factor; this is sometimes in association with partial lipodystrophy *(7,8)*.

6. Increased susceptibility to Neisserial infections. The clearest association between C deficiency and disease is with susceptibility to Neisserial infections. Neisserial disease, particularly meningococcal disease, has been described in patients with genetic deficiencies of classical or alternative complement components, as well as terminal components *(9)*, and tests of complement function in affected patients may need to include all pathways.

Nevertheless, it is with terminal component and properdin deficiencies that meningococcal infections are most closely associated *(10,11)*. N. meningitidis infections remain a serious cause of disease and death among normal, complement-sufficient children and young adults both in Western and developing countries, and a proportion of complement-deficient individuals give no history of Neisserial disease and appear healthy. However, there are certain factors that should alert clinicians to the possibility of C deficiency. Terminal component deficiencies typically are diagnosed in older children, adolescents, or young adults presenting with recurrent episodes of meningococcal disease, or with disease caused by organisms with uncommon polysaccharide capsule antigens, such as serogroups Y or W135 *(12)*. All cases of recurrent disease have had a primary episode, sometimes in childhood, and the prevalence of complement deficiency in primary cases to a certain extent varies with incidence of meningococcal disease in the community. It is not yet fully understood why susceptible complement-deficient individuals frequently do not become infected until they are in their teens or early adulthood. The serogroups identified in the disease strains are just one of several markers used in investigation of the epidemiology of N. meningitidis infections. Studies of these markers indicate that terminal component deficient individuals are susceptible to infections that would not cause disease in healthy individuals *(13)*. This, of course, does not protect them from infection with common, or epidemic, strains; so the association with uncommon serogroups is far from absolute.

Properdin deficiency is X linked *(11)* and, unlike terminal component deficiencies, antibodies produced after vaccination or primary infections are able to mediate bactericidal protective activity. The deficiency is familial and most commonly diagnosed in males presenting with severe primary meningococcal infections, often caused by rare serogroups *(3,11)*. It is important to remember that this deficiency is not detected by classical complement screening, and

Table 1
Disease Presentations Associated with Complement Deficiency

Disease presentation	Responsible or associated complement deficiency	Functional defect	Ref.
Immune complex (IC) diseases, SLE with and without glomerulonephritis (GN).	Deficiency of all classical pathway components, most severe with C1q. Occurs also with C3 deficiency.	Impaired IC handling.	*1,2,3*
Systemic infections with encapsulated bacteria (including *Neisseria*)	Deficiencies of classical components or C3 deficiency syndromes; often together with the other manifestations of deficiency.	Impaired opsonophagocytosis and bactericidal activity.	*3*
Hereditary angioedema	C1 inhibitor deficiency	(*see* chapter 12)	(*see* chapter 12)
Severe and often repeated pyogenic infections.	Deficiency of C3 either primary, or secondary to fH or fl deficiency, or the presence of C3 or C4 nephritic factors. Can also occur with classical component deficiency.	Impaired complement activation and impaired opsonophagocytosis.	*3,4,5,6,7*
Mesangiocapillary glomerulonephritis (MCGN) or membraoproliferative glomerulonephritis (MPGN)	C3 deficiency classically caused by C3 or C4 nephritic factors, also fH deficiency.	Hypocomplementaemia caused by autoantibodies to convertases.	*5,7,8*

Neisserial infections, usually recurrent meningococcal infections. Occasionally disseminated N. gonorrhoea infections.	Terminal component (C5–8) deficiencies; the association is less strong in patients with C9 deficiency. Patients are usually otherwise healthy.	Impaired serum bactericidal activity.	9,10
Severe, occasionally repeated, N. meningitidis infections in males.	Properdin deficiency or Factor D deficiency (very rare, not sex linked)	Impaired or absent alternative complement pathway function.	3,11,26
Repeated infections in early childhood.	Low serum levels/deficiency MBL.	Impaired MBL complement activation, impaired opsonophagocytosis	14
Leukocyte adhesion deficiency with severe bacterial infections in infancy.	CR3 deficiency (rare).	Defect in iC3b mediated opsonophagocytosis, impaired leukocyte adhesion.	15
Paroxysmal nocturnal hemoglobinuria	Rare acquired clonal disorder of hemopoietic cells which lack glycosylphosphatidylinositol (GPI) membrane anchored complement control proteins [(DAF) and CD59].	Impaired regulation of deposition of C3b and C8 on host red blood cells and other	2,16

occasionally may not be evident on AP screening so that the diagnosis may need specific assays.

The importance of diagnosis of terminal component or properdin deficiency in index cases, and their family members, is that there are strong indications that patients benefit from the quadrivalent meningococcal vaccine (A, C, Y, W135).

7. Other disease associations. Other deficiencies of the complement system responsible for disease syndromes include deficiency of mannan-binding lectin (MBL) *(14)*, which may be responsible for susceptibility to repeated respiratory infections in early childhood, deficiencies of complement receptors causing leukocyte adhesion abnormalities *(15)*, and deficiency of cellular membrane control proteins, such as decay-acceleration factor (DAF) *(16)* and CD59 *(2)*. These deficiencies are interesting but are, apart from MBL deficiency, rare; moreover, as they will not be identified by the complement screening methods discussed here, I will not go into further detail.

2. Materials

2.1. Samples

1. Serum samples. Samples for the functional hemolytic assays need to be fresh, that is serum should be separated on the day of venepuncture and used the same day, or stored at $-80°C$. This is probably the single most difficult, yet important, step in organizing complement screening, because if the cold chain is broken, the results become impossible to interpret correctly. Immunochemical assays of individual components are not as vulnerable as the assays that depend on functional activity, and the samples can be stored at $-10°C$. Testing for genetic deficiencies of complement is best avoided on serum samples obtained during the acute phase of an infectious disease such as meningococcal septicaemia, as complement activation can occur, leading to temporary hypocomplementemia *(17)*. This finding can be an important indicator of serious disease, but would also interfere with identification of a predisposing genetically determined complement deficiency.

2. DNA samples for molecular biology tests are best prepared from whole blood, with ethylenediaminetetraacetic acid (EDTA) as the anticoagulant, by well-established methods *(18)*.

2.2. Buffers and Other Reagents

1. Complement fixation diluent (CFD) should be used for complement assays where Ca^{++} and Mg^{++} are required. It can be obtained in preprepared tablet form from Oxoid (Basingstoke, UK). Briefly, it has a pH 7.2 and contains: 3.1 mM diethyl barbituric acid, 0.9 mM sodium barbitone, 145 mM NaCl, 0.83 mM $MgCl_2$, 0.25 mM $CaCl_2$. It can be stored as X 5 stock with 0.005 M sodium azide. CFD/gelatin contains 1 g/L gelatin and is used when protein concentrations are low.

2. Phosphate-buffered saline (PBS) contains: 137 mM NaCl, 2.7 mM KCl, 1.5 mM KH_2PO_4, 8.1 mM Na_2HPO_4. Sodium azide can be added if required. PBS/EDTA is PBS containing 10 mM EDTA.

3. Alternative pathway buffer (AP buffer) is PBS containing 100 m*M* ethylene gly-col-*bis*(β-aminoethyl ether)N,N',N'-tetraacetic acid (EGTA) and 7 m*M* MgCl$_2$; resulting in chelation of Ca^{++}, but not Mg^{++}, and providing additional Mg^{++}. This prevents complement activation via the CP and facilitates complement activation via the AP in agarose gels.
4. Agarose: use an agarose that has a low melting point as plates are easier to pour. We now use Sigma Type IV-A. A 2% stock solution in water is made.
5. Zymosan A (zymosan) from *Saccharomyces cerevisiae* and antrypol are obtained from Sigma.

2.3. Equipment

2.3.1. For agarose gels:

1. Suitable glass plates. Size depends on the number of samples, every sample requires an area of approximately 2.25 cm^2 (1.5 × 1.5 cm). Volumes given in the method are for 8 × 8-cm plates. Volumes can be adjusted so that the final depth of the gel is ~1–1.5 mm.
2. 56°C water bath.
3. 45°C water bath.
4. Plate warmer.
5. Level table.
6. 37°C incubator.
7. Well cutters to produce holes approx 3 mm diameter [if not available, large (size 15) hypodermic needles cut to have a straight aperture can be used]. If single holes are to be punched, a grid should be placed under the gel so that the holes can be distributed evenly.

2.4. Target Cells

2.4.1. Sheep Red Blood Cells (E) and Guinea Pig Red Blood Cells (GPE)

Red blood cells are best obtained by taking blood into a chelating anticoagulant such as acid-citrate dextrose (ACD). Sterile blood samples can be used up to 3 wk after venesection if stored in Alsever's solution at 4°C. When required, E or GPE in Alsever's solution are prepared by washing the cells three times in PBS, once in the final buffer (CFD or AP buffer) and suspending in the final buffer at 10%. These washed suspensions can be kept 1–2 d at 4°C.

2.4.2. Red Cell Intermediates (see Chapter 4 for Detailed Protocols)

1. E are sensitized by treating with antisheep red blood cell (RBC) antibody (A) to produce EA. Traditionally, rabbit antisheep RBC antibody from Wellcome Diagnostics was used, but recently we have used rabbit hemolytic serum (antisheep RBC antibody) from TCS Microbiology, Buckingham, UK; both are satisfactory. Certain laboratories produce their own monoclonal antibodies (MAbs) to sheep RBC and these can be very potent *(19)*. The maximum antibody that does not cause

hemagglutination should be used; this is determined by titration in round-bottom microtiter plates. Commercial antisera generally need to be diluted about 1/200.

2. EAC [EA14(2)3] are sheep RBC sensitized with antibody and the early human complement components [C2 function is labile and therefore written as (2)]. These are made by treating EA with serum containing the early human complement components, but not the lytic terminal components (Chapter 4). The serum is prepared by treating human serum with yeast or zymosan and is referred to as R3 by Harrison and Lachmann *(19)*; however, as they point out, it is important to realize that it is not completely C3 depleted. I will refer to it here as zymosan-treated serum.

2.5. Complement Deficient/Depleted Reagents

These are required for specific protocols for individual component deficiencies. There are many such protocols, but most individual components are now most readily measured using antibody assays, so that the full range of possible functional assays is not given here. However, the reagents for the human terminal complement component assays are detailed below. Most other protocols are covered in detail in Chapter 4 or in Harrison and Lachmann *(19)*.

1. For C5 assays, serum from certain inbred mouse strains such as the A/J strain that are C5 deficient (C5D) can be used (Chapter 17).
2. C6 assays require C6 deficient (C6D) rabbit serum, which is obtainable from colonies of C6D rabbits (Chapter 17).
3. C7 functional assay requires C5b6 euglobulin. This should be prepared from serum obtained from individuals with acute phase inflammation (postpartum serum is very good, but serum from patients following sports injuries or surgery is also good) (Chapter 2). Alternatively, use human C7 deficient (C7D) serum or C7 depleted serum.
4. Assays for complete analysis of C8 functional activity use human C8β deficient (C8βD) and C8α-γ deficient (C8α-γ) deficient serum *(20)*. However, C8βD is more common in Western countries and a patient lacking a complete C8 molecule can be analyzed for C8βD with this serum.
5. Assays for C9 functional activity require C9 depleted serum (Chapter 4) *(21)*.

3. Methods
3.1. Preparation of Zymosan-Treated Serum

1. To 1 mL human serum, add 10 mg of untreated zymosan (Sigma), incubate at 37°C for 45 min with gentle shaking.
2. Centrifuge to remove the zymosan. Store in aliquots at –80°C.

3.2. Preparation of EAC

1. Dilute 1 mL 10% EA to 10 mL 1% EA in prewarmed CFD, keep at 37°C, and add 1 mL zymosan-treated serum.

2. Shake the mixture gently for 20 s and then add 0.04 mL Antrypol (50 mg/mL in CFD).

3. After gently shaking at 37°C for 5 min, the red cells are washed in cold (4–8°C) CFD, if possible, using a refrigerated centrifuge, and resuspended at 10%. The cells, EAC14(2)3(hu)(Antrypol) (abbreviated to EAC when used in the context of complement assays), now have the early components on their surface.

4. EAC should be kept on ice and used the same day.

3.3. Making Prepared Yeast

1. Suspend 500 g baker's yeast (young and free of mycelia) in 2 L PBS and autoclave 30 min at 120°C.

2. Wash the yeast in PBS, using centrifugation, until the supernatant is clear.

3. Suspend the washed yeast in 250 mL PBS, and, working in a fume hood, add 1.7 mL 2-mercaptoethanol and stir at 37°C for 2 h.

4. Centrifuge, wash with PBS, and recentrifuge. Monitor pH and resuspend yeast in 500 mL 0.04 M Na$_2$ (or K$_2$) phosphate, 0.145 M NaCl, 0.02 M iodoacetamide at pH 7.2.

5. Stir at 22°C for 2 h to alkylate the yeast. Adjust the pH to 7.2 with 1 M NaOH if necessary.

6. Centrifuge and wash with PBS three times and suspend finally in 2 L PBS. Autoclave 30 min at 120°C and wash in PBS until supernatant is clear.

7. Store in 1 L CFD azide at 4°C. The prepared yeast is titrated against guinea-pig serum to determine the minimum amount that removes all complement activity after incubation at 37°C for 45 min (generally 0.2 mL yeast per mL serum) and this quantity is used for treating human serum (*see* below).

3.4. Preparation of C5b6 Euglobulin

1. To fresh acute phase serum, add yeast suspension (amount from the titration described above) and incubate 37°C for 45 min or overnight at 22°C.

2. Centrifuge to remove yeast and dialyze, at 4°C, against 0.02 M phosphate pH 5.4.

3. Collect the precipitate (C56 euglobulin) by centrifugation and dissolve in original volume PBS.

4. C5b6 euglobulin for C7 assays should lack C7, but should have C8 and C9 present; it is not a pure preparation.

3.5. Hemolytic Classical Complement Assay in Agarose Gels (CP Gels)

1. Melt 2% agarose stock (most conveniently in a microwave oven although immersion in boiling water will suffice) and, using a warm pipet, pipet 3.5-mL aliquots into universal containers (one for each gel), keep at 56°C.

2. Warm 2 X CFD to 56°C and add 2.8 mL to each bottle of melted agarose. Mix well.

3. Transfer one bottle with diluted agarose to 45°C and allow to cool to this temperature.

4. Warm glass plate (to about 35°C) to ease pouring the gel, and place on level tray.

5. Add 0.7-mL 10% EA to the diluted agarose at 45°C, mix carefully and quickly.
6. Pour the mixture evenly onto the level plate. The mixture should go to the edges, but not run over the edges. It may help to have the plate slightly raised so that the edges do not touch anything.
7. Remove bubbles by touching them with a gloved finger.
8. Cool plate to 4°C, punch holes at least 1.5 cm apart.
9. Fill wells with a measured serum volumes (5 up to 8 μL). If crude quantitation of complement activity is required include a normal human serum (NHS) standard, and NHS diluted 1/2 and 1/4 on each individual plate. The size of rings depends on factors such as gel thickness, and NHS standards are needed on each plate.
10. Incubate overnight at 4°C, examine plate before transferring to 37°C. Incubate at 37°C for 1–2 h.
11. Measure the diameters of the rings of lysis and calculate the areas. Areas of lysis can be read after photography, or after making direct photographic prints, but this is optional.
12. A crude standard curve is drawn by plotting % concentration NHS vs area of lysis (diameter squared can also be used). This allows calculation of % normal activity in test samples. However, this calculation is, at best, semiquantitative and is only a guide (*see* **Note 1**). This is because diluting NHS means that all components are diluted equally, whereas patients with deficiencies usually have markedly reduced activity of just one component.

3.6. Alternative Complement Pathway Activity in Agarose Gels (AP Gels).

The procedure is essentially the same as for the CP gels except that 2 X AP buffer and 10% GPE are used to make the gels.

1. Prepare gels as described in **steps 1–8** in **Subheading 3.5.**, substituting 2XAP buffer for 2X CFD and 10% GPE for 10% EA.
2. Place serum samples and controls in wells as described in **step 9** in **Subheading 3.5.**
3. Incubate either overnight at 37°C, or overnight at 22°C and 2 h at 37°C.
4. Measure the diameters of the rings of lysis and calculate the areas. Areas of lysis can be read after photography, or after making direct photographic prints but this is optional.
5. Calculate percentage activity in test sera essentially as described in **step 12, Subheading 3.5.** The size of the areas of lysis often varies widely. *See* **Note 2**.

For interpretation, *see* **Note 1**.

3.7. Double Diffusion Hemolytic Assay for Individual Component Deficiencies

This method was developed because it allows simultaneous investigation of several specific complement deficiencies *(22)*. The method has the advantage of simultaneously testing for several deficiencies and requiring very small amounts of the deficient reagents. It has been used in tests for deficiencies of C2 and all terminal complement components, except C8α-γ, and as more sera from complement deficient patients or animals, become available it will be usable on a wider scale. It is very efficient and informative.

1. CP plates are prepared as described in **Subheading 3.5.** above. If only one set of holes is required a much smaller plate will suffice (the original reference details making the plate with EA treated with zymosan treated human serum [the red cell intermediate EA14(2)3), but this is not necessary and EA also work well].
2. Five 3-mm holes are punched in the gel (**Fig. 1**), the centers of the four peripheral holes being approx 9 mm from the center of the central hole. Distance is important as the assay depends on diffusion.
3. Test serum (7 μL) is inoculated into the central hole and 7 μL of human serum samples, each known to be deficient in one of the components under test, are inoculated into the peripheral wells.
4. The gel is incubated overnight at 4°C and then at 37°C for approx 6 h. The samples in the peripheral wells in **Fig. 1** are C5D, C6D, C7D, and C8βD sera. The figure shows that double lines of hemolysis developed between the test sample and C5D, C6D, and C8βD sera showing that the test sample is not deficient in C5, C6, or C8β. However, no lines developed between the test sample and the C7D sample showing the test sample was also a C7D sample.

See **Note 4**.

3.8. Special Methods for Individual Components

The CP and AP screening assays described above will, if properly performed, diagnose the presence of most genetic deficiencies of complement, and indicate whether further tests of the CP components, C3 or AP components, or terminal components should be done (*see* **Note 3**).

3.8.1. Immunochemical Assays for Individual Components

Immunochemical assays are frequently the most suitable for detecting deficiencies of individuals components. Routine immunochemical assays, such a nephelometry, are particularly appropriate for measuring C3 and C4 levels and these determinations should preferably be done together with the initial complement screening. Low levels of these components are more often caused by complement consumption secondary to conditions, such as immune complex disease, acute angioedema, or severe systemic infections, than to genetic deficiencies, and this information is required for interpretation of functional assays. C2 levels are also valuable for detecting consumption, but I have encountered difficulties with commercially available C2 immunochemical assays.

Monoclonal antibodies to most complement proteins have been developed so that sensitive assays can be used for almost all individual components. However,

Table 2
Outline of Steps in Screening for Complement Deficiency

	Initial screen. Classical pathway functional assay Alternative pathway functional assay C3 and C4 levels by immunochemical testing		
Results (Abnormal)	(A) Absent or very low CP activity. AP activity present.	(B) CP activity normal. AP activity absent or very low	(C) Absent CP and absent AP activity.
Interpretation	One or more CP components or C3 consumed or deficient. Terminal pathway normal.	CP components normal. AP can be normal with low activity (*see* **Subheadings 4.2, 4.3.,** and **4.5.**). Suspect AP deficiency if history suggestive. Also suspect C9 deficiency (*see* **Subheading 4.4.**)	Terminal component deficiency. C3 deficiency.

Further tests	Determine CP component and C3 levels.	(i) Determine C3 level. (ii) Determine properdin and factor D levels if indicated. (iii) Determine C9 level.	Determine terminal component levels and C3 levels. Check history.
Results	(i) C2, C4, and/or C3 low and the presence of other disease that produces complement activation. (ii) Absence of an individual CP component or C3, with or without disease associated with CP deficiency.	(i) C3 very low or absent. Check history. (ii) Properdin or fD absent. (iii) C9 low or absent.	(i) Absent or very low activity of a terminal component. (ii) Absent or very low C3 levels.
Interpretation	(i) Activation and utilization of CP. (ii) CP component or C3 deficiency.	(i) C3 deficiency syndrome. (ii) Properdin or fD deficiency. (iii) C9 deficiency.	(i) Terminal component deficiency. (ii) C3 deficiency syndrome.

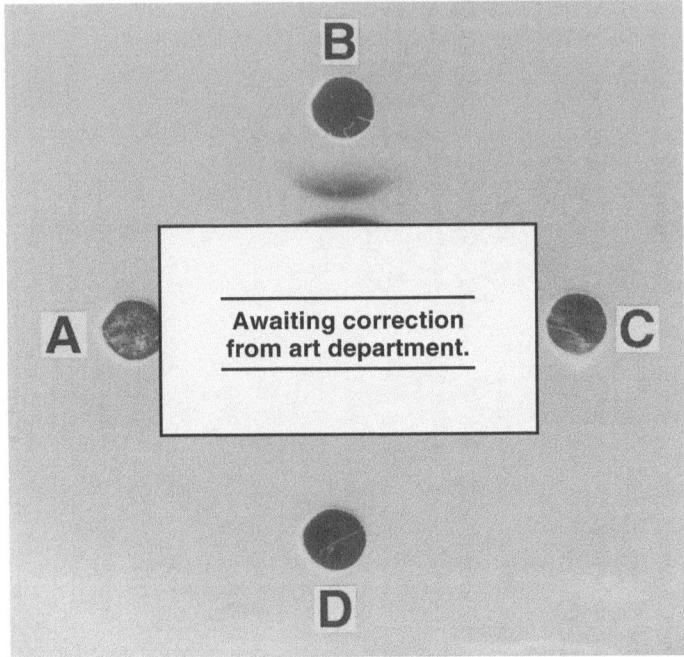

Fig. 1. The hemolytic gel contains sensitized sheep red cells, EA14(2)3 (EAC). The test sample (7 µL was inoculated into well **E**, and 7 µL each of human C6 deficient, C5 deficient, C8β deficient, and C7 deficient sera were inoculated into wells **A**, **B**, **C**, and **D**, respectively. Incubation was at 4°C for 17 h and at 37°C for 6 h. Two bands of lysis developed between the test sample and wells **A**, **B**, and **C** (the band nearest to well **A** is very faint), but no lines developed between the test sample and well **D**. This indicates that, like the serum in well **D**, the test sample is C7 deficient. [*from* Egan, L., Orren, A., Doherty, J., Würzner, R., and McCarthy, C. F. (1994) Hereditary deficiency of the seventh component of complement and recurrent meningococcal infection: investigations in an Irish family using a novel hemolytic screening assay for complement activity and C7 M/N allotyping. *Epidemiol. Infect.* **113**, 275–281. *Cambridge Univ. Press. With permission of the publisher.*]

these are not all commercially available. One important consequence of sensitive immunochemical assays such as ELISAs is that it is now possible to distinguish subtotal from complete deficiency of many components *(23)*. *See* **Notes 5** and **6**.

3.8.2. Investigations of Patients With Very Low or Absent C3 Levels

This can be caused by primary C3 deficiency, deficiencies of fI or fH or the presence of a C3 nephritic factor (NeF). Patients with primary C3D have low

C3 and low or absent C3 breakdown products. Patients with fI or fH deficiency have evidence of increased C3 consumption and the majority of the C3 they do have tends to be in the form of C3b; further C3 breakdown products (C3d, C3d,g) are absent *(24,25)*. Elegant electrophoretic methods can be used to make these diagnoses *(25)*, although today it may be more direct and simpler to initially measure fI and fH levels immunochemically. In patients with C3 (NeF), the NeF stabilizes the C3 convertase C3bBb, thus leading to the production of more C3b and further C3 breakdown continues; the C3b breakdown frequently results in increased levels of breakdown products C3d and C3d,g. When normal serum is exposed to NeF containing serum, the NeF serum causes C3 conversion in the normal serum *(19)*. This can be determined by crossed immunoelectrophoresis *(25)* or radioimmunoelectrophoresis *(24)*. The importance of NeF for complement deficiency is that it is a cause, frequently temporary, of what can be very low C3 levels, and the accompanying susceptibility to bacterial infections.

3.8.3. Deficiencies of Regulator Proteins

Most deficiencies of circulating complement regulator proteins affect the levels of other components.

1. The majority of cases of C1 esterase inhibitor deficiency present as angioedema (Chapter 12). Deficiencies of C1 esterase inhibitor are usually determined by means of immunochemical assays, however, symptoms also occur in people with normal antigenic levels of defective proteins. This can be tested for by specific assays for the function of the enzyme inhibitor. The acute attacks of angioedema that occur in these patients lead to consumption of classical complement components and the resultant low levels of C4 and C2 may aid diagnosis.
2. fI and fH deficiency usually present as secondary C3D.
3. Properdin. Deficiency is usually, but not always detected in AP plates (A. Sjöholm personal communication). Therefore, if there are clinical reasons to suspect properdin deficiency, specific properdin assays should be employed. Usually antibody assays such as enzyme-linked immunosorbent assay (ELISA) *(26)* will make the diagnosis, but properdin deficiency arising because of the presence of a dysfunctional protein would be missed by such assays, and required a special functional assay in which fI and fH were removed from the serum prior to a hemolytic assay *(27)*.

3.8.4. Deficiencies of Terminal Complement Components

Deficiency of a terminal complement component can be tentatively diagnosed when a fresh serum sample produces no lysis on either the CP or AP plate and C3 levels are not markedly depressed. Monoclonal antibodies to the terminal components are now available and these antibodies in ELISA assays,

or other immunochemical assays, can be used to identify the deficient component. Alternatively, functional assays can be used *(19)*. These functional assays
are usually adaptations of the CP or AP assays and all require a source of serum
depleted of, or deficient in, the component under study to ensure that the component under study is limiting. Therefore the choice of whether antibody or
functional assays are used will probably depend on whether deficient/depleted
sera or antibodies are available.

3.8.5. Deficiency of MBL

Deficiency of MBL will not be diagnosed by the use of the CP and AP
plates. It is associated particularly with respiratory infections in early childhood *(14)* and needs to be tested for independently using antibody assays if it is
being considered.

3.9. Determination of the Genetic Molecular Defects Underlying Complement Deficiency

These methods determine the DNA molecular defects that are responsible
for particular deficiencies. However, the molecular defects leading to complement deficiencies are very heterogeneous and many have not yet been identified *(28)*. This means that DNA molecular methods are generally not useful as
initial screening methods. Nevertheless, as we grow to know more of the
genetic defects responsible for deficiencies, and as screening for DNA defects
becomes more widely used, DNA methods will be used for complement deficiency screening. These methods are particularly appropriate in certain geo-
graphical areas where particular complement deficiencies, caused by segregation
of particular defects within the community, occur with unusually high frequencies *(29)*. I will not provide sequencing, restriction fragment length polymorphism (RFLP), or other basic methods for detection of the molecular defects
here; there are very many references to these methods. However, a point of
particular practical interest is that investigators who have identified a complement deficiency in a serum sample can proceed to extract sufficient DNA from
that serum (or plasma) sample for various polymerase chain reaction (PCR)
investigations *(30,31)* and thus to identification of the molecular defect. We
have used the QIAamp Viral RNA kit (the old name was QIAMP HCV kit)
(QIAGEN GmbH, Hilden, Germany) for DNA extraction form serum/plasma,
which we found better than the one used by Fowke et al. *(30)*.

The method works well on stored samples that have been frozen and thawed,
as well as on fresh samples. The kit is intended primarily for the isolation of
viral RNA from small amounts of body fluids, but simultaneously harvests any
DNA present in the sample. Full instructions are provided with the kit which
should be followed, except for the use of DNAse as this step must definitely

be left out. Sufficient DNA for up to 10 PCR experiments is obtainable from 250–300 μL of serum. The amount of DNA present in serum varies and up to 40 cycles per PCR are sometimes necessary to obtain adequate material.

4. Notes

1. The hemolytic tests in agarose gels are screening tests. Although they can be used to estimate levels of complement function they will not give accurate quantitative results. The CP assay depends on antibody-sensitized sheep red cells and is more sensitive than the AP assay. Absent or low CP activity with AP activity present does indicate low or absent levels of a CP component and levels of the individual components should be tested. However, caution should be exercised in that C2 in particular is labile, and tests do require the use of fresh samples. Also, samples taken during an acute phase of an infectious or autoimmune disease frequently have low levels caused by consumption and are also more labile than samples from healthy individuals.

2. Interpretation of AP plates is difficult. The range of size of AP hemolytic rings is wide and small rings do not necessarily indicate genetic AP component deficiency. One of the main uses of the AP test in screening is that it allows the differentiation of terminal component deficiency from classical pathway deficiency, and small rings in both CP and AP plates suggest low C3 levels.

3. Results of hemolytic tests for complement deficiency sometimes depend on the sensitivity of the red cell intermediates to complement lysis. Problems arise if the hemolytic tests are too sensitive to detect the reduced AP function in properdin deficiency, and reduced CP function in C9 deficiency. Some workers use rabbit erythrocytes (RE) instead of GPE in AP assays *(32)*. However, comparison of GPE and RE for these assays showed RE to be exquisitely sensitive and to allow partial lysis with C5D and C7D sera *(33)* that could lead to incorrect interpretation of complement screening.

4. Special precautions need to be taken in looking for C9 deficiency because, depending on the conditions of the assay, C9 is not an absolute requirement for hemolysis *(34)* so that C9D probably will not be detected with the routine CP assay. In Japan, a specific CP assay in which C9D sera produced partial hemolysis, was used to screen for C9 deficiency, and appeared to work well *(32)*. The final confirmations of the C9 deficiencies were made immunochemically. Alternatively, an interesting use of AP tests that my colleagues and I observed was in the detection of C9 deficiency. Two subjects were tested by the CP and AP screening tests given here and both had absent AP activity and present, although reduced, CP activity. Further investigation *(33,35)* showed both patients to be C9 deficient, one almost total deficiency and one subtotal deficiency. Thus, the sensitive CP plates were not suitable for detecting the reduced terminal complement activity resulting from C9 deficiency, but the less-sensitive AP plates were. I have not used the CP assay reported by Fukumori et al. *(32)*, but suggest that anyone wishing to screen for C9 deficiency should look at that method, or use AP assays as well as CP assays, or use a specific C9 immunochemical assay. It seems

very probable that, in countries outside the Far East, C9 deficiency has a higher prevalence than is at present recognized .

5. The markedly reduced activity of CP function found in patients with a CP, or terminal component subtotal deficiency will be detected on the CP plate. However, this screening assay does not provide good quantitative results and immunochemical analysis is preferable for quantitative investigations of subtotal deficiencies.

6. The functional assays described here have the advantage that they detect abnormalities brought about by dysfunctional proteins, whereas immunochemical assays would usually fail to detect them. There are several complement deficiencies where it has been shown to be particularly important to be able to detect dysfunctional proteins, these include C1 esterase inhibitor deficiency (*see* Chapter 12) and properdin deficiency *(27)*, and for both of these special functional assays need to carried out.

References

1. Schifferli, J. A., Ng, Y. C., and Peters, D. K. (1986) The role of complement and its receptor in the elimination of immune complexes. *N. Engl. J. Med.* **315,** 488–495.
2. Morgan, B. P. and Walport, M. J. (1991) Complement deficiency and disease. *Immunol. Today* **12,** 301–306.
3. Figueroa, J. E. and Densen, P. (1991) Infectious diseases associated with complement deficiencies. *Clin. Microbiol. Rev.* **4,** 359–395.
4. Botto, M., Fong, K. Y., So, A. K., Rudge, A., and Walport, M. J. (1990) Molecular basis of hereditary C3 deficiency. *J. Clin. Invest.* **86,** 1158–1163.
5. Thompson, R. A. and Winterborn, M. H. (1981) Hypocomplementaemia due to a genetic deficiency of beta 1H globulin. *Clin. Exp. Immunol.* **46,** 110–119 .
6. Vyse, T. J., Späth, P. J., Davies, K. A., Morley, B. J., Philippe, P., Athanassiou, P., et al. (1994) Hereditary complement factor I deficiency. *Q. J. Med.* **87,** 385–401.
7. Skattum, L., Martensson, U., and Sjöholm, A. G. (1997) Hypocomplementaemia caused by C3 nephritic factors (C3 NeF): clinical findings and the coincidence of C3 NeF type II with anti-C1q autoantibodies. *J. Intern. Med.* **242,** 455–464.
8. Levy, Y., George, J., Yona, E., and Shoenfeld, Y. (1998) Partial lipodystrophy, mesangiocapillary glomerulonephritis, and complement dysregulation. An autoimmune phenomenon. *Immunol. Res.* **18,** 55–60.
9. Ross, S. C. and Densen, P. (1984) Complement deficiency status and infection: epidemiology, pathogenesis and consequences of Neisserial and other infections in an immune deficiency. *Medicine* (Baltimore) **63,** 243–273.
10. Würzner, R., Orren, A., and Lachmann, P. J. (1992). Inherited deficiencies of the terminal components of human complement. *Immunodef. Rev.* **3,** 123–147.
11. Sjöholm, A. G., Braconier, J.-H., and Söderström, C. (1982) Properdin deficiency in a family with fulminant meningococcal infections. *Clin. Exp. Immunol.* **50,** 291–297.
12. Fijen, C. A., Kuijper, E. J., Hannema, A. J., Sjöholm, A. G., and van-Putten, J. P. (1989) Complement deficiencies in patients over ten years old with meningococcal disease due to uncommon serogroups. *Lancet ii* **(8663),** 585–588.

13. Orren, A., Caugant, D. A., Fijen, C. A. P., Dankert, J., Van Schalkwyk, E. J., Poolman, J. T., and Coetzee, G. J. (1994) Characterization of strains of *Neisseria meningitidis* recovered from complement -normal and complement-deficient patients in the Cape, South Africa. *J. Clin. Microbiol.* **32,** 2185–2191.

14. Super, M., Thiel, S., Lu, J., Levinsky, R. J., and Turner, M. W. (1989) Association of low levels of mannan-binding protein with a common defect of opsonisation. *Lancet ii* **(8674),** 1236–1239.

15. Kishimoto, T. K., Hollander, N., Roberts, T. M., Anderson, D. C., and Springer, T. A. (1987) Heterogeneous mutations in the beta subunit common to the LFA-1, Mac-1, and p150,95 glycoproteins cause leukocyte adhesion deficiency. *Cell* **50,** 193–202.

16. Rosse, W. F. (1997) Paroxysmal Nocturnal Haemoglobinuria (PNH) as a molecular disease. *Medicine* (Baltimore) **76,** 63–93.

17. Brandtzaeg, P., Mollnes, T. E., and Kierulf, P. (1989) Complement activation and Endotoxin levels in Meningococcal Disease. *J. Infect. Dis.* **160,** 58–65.

18. Jeffreys, A. J. (1979) DNA sequence variants in the $^G\gamma$-, $^A\gamma$-, δ- and β- globin genes of man. *Cell* **18,** 1–10.

19. Harrison, R. A. and Lachmann, P. J. (1986) Complement technology, in *Handbook of Experimental Immunology,* vol. 1 (Weir, D. M., ed.), Blackwell Scientific, Oxford, pp. 39.1–39.49.

20. Tedesco, F., Densen, P., Villa, M. A., Petersen, B. H., and Sirchia, G. (1983) Two types of dysfunctional eighth component of complement (C8) molecules in C8 deficiency in man. Reconstitution of normal C8 from the mixture of two abnormal molecules. *J. Clin. Invest.* **71,** 183–191.

21. Morgan, B. P., Daw, R. A., Siddle, K., Luzio, J. P., and Campbell, A. K. (1983) Immunoaffinity purification of human complement C9 using monoclonal antibodies. *J. Immunol. Methods* **64,** 269–281.

22. Egan, L., Orren, A., Doherty, J., Würzner, R., and McCarthy, C. F. (1994) Hereditary deficiency of the seventh component of complement and recurrent meningococcal infection: investigations in an Irish family using a novel haemolytic screening assay for complement activity and C7 M/N allotyping. *Epidemiol. Infect.* **113,** 275–281.

23. Orren, A., Würzner, R., Potter, P. C., Fernie, B. A., Coetzee, S., Morgan, B. P., and Lachmann, P. J. (1991) Properties of a low molecular weight complement component C6 found in human subjects with subtotal C6 deficiency. *Immunology* **75,** 10–16.

24. Lachmann, P. J., Pangburn, M. K., and Oldroyd, R. G. (1982) Breakdown of C3 after complement activation: identification of a new fragment, C3g, using monoclonal antibodies. *J. Exp. Med.* **156,** 205–216.

25. Teisner, B., Brandslund, I., Folkersen, J., Rasmussen, J. M., Poulsen, L. O., and Svehag, S.-E. (1984) Factor I deficiency and C3 nephritic factor: Immunochemical findings and association with *Neisseria meningitidis* infection in two patients. *Scand. J. Immunol.* **20,** 291–297.

26. Densen, P., Weiler, J. M., Griffiss, J. M., and Hoffmann, L. G. (1987) Familial properdin deficiency and fatal meningococcaemia. Correction of the bactericidal defect by vaccination. *N. Engl. J. Med.* **316,** 922–926.

27. Sjöholm, A. G., Kuijper, E. J., Tijssen, C. C., Jansz, A., Bol, P., Spanjaard, L., and Zanen, H. C. (1988) Dysfunctional properdin in a Dutch family with meningococcal disease. *N. Eng. J. Med.* **319,** 33–37.

28. Würzner, R., Witzel-Schlömp, K., Tokunaga, K., Fernie, B. A., Hobart, M. J., and Orren, A. (1998) Reference Typing Report for Complement Components C6, C7 and C9, including mutations leading to deficiencies. *Exp. Clin. Immunogenet.*, **15,** 268–285.

29. Hobart, M. J., Fernie, B. A., Fijen, K. A. P. M., and Orren, A. (1998) The molecular bases of C6 deficiency in the Western Cape, South Africa. *Hum. Genet.*, **103,** 506–512.

30. Fowke, K. R., Plummer, F. A., and Simonsen, J. N. (1995) Genetic analysis of human DNA recovered from minute amounts of serum or plasma. *J. Immunol. Meth.*, **180,** 45–51.

31. Fernie, B. A., Finlay, A., Price, D., Chan, E., Orren, A., Joysey, V. C., et al. (1996) Complement C6 and C7 DNA polymorphisms analysed by PCR in seven ethnic groups and characterisation of the C6 MspI RFLP. *Exp. Clin. Immunogenet.* **13,** 92–103.

32. Fukumori, Y., Yoshimura, K., Ohnoki, S., Yamaguchi, H., Akagaki, Y. and Inai, S. (1989) A high incidence of C9 deficiency among healthy blood donors in Osaka, Japan. *Int. Immunol.* **1,** 85–89.

33. Hobart, M. J., Fernie, B. A., Würzner, R., Oldroyd, R. G., Harrison, R. A., Joysey, V., and Lachmann, P. J. (1997) Difficulties in the ascertainment of C9 deficiency: lessons to be drawn from a compound heterozygote C9-deficient subject. *Clin. Exp. Immunol.* **108,** 500–506.

34. Stolfi, R. L. (1968) Immune lytic transformation: a state of irreversible damage generated as a result of the reaction of the eighth component in guinea pig complement system. *J. Immunol.* **100,** 46–54.

35. Witzel-Schlömp, K., Hobart, M. J., Fernie, B. A., Orren, A., Würzner, R., Rittner, C., et al. (1998) Heterogeneity in the genetic basis of human complement C9 deficiency. *Immunogenetics* **34,** 144–147.

12

C1-Inhibitor: *Antigenic and Functional Analysis*

C. Erik Hack

1. Introduction

The classical pathway of complement is activated in various diseases. A main inhibitor of this pathway is the serpin C1-inhibitor (C1-Inh), which regulates activation at the level of C1. C1-Inh also inhibits the mannan binding lectin (MBL) associated serine proteinases-1 and -2. A deficiency of C1-Inh can result from a heterozygous genetic defect and is associated with a disease characterized by the occurrence of bouts of angio-edema (hereditary angioedema or HAE). A deficiency of C1-Inh can also be acquired (AAE) because of enhanced consumption of C1-Inh. AAE is often associated with underlying diseases, in particular, B-cell malignancies. A diagnosis of C1-Inh deficiency can be made by analysis of plasma levels of functional C1-Inh. HAE and AAE can be treated amongst others by substitution with C1-Inh purified from pooled plasma from healthy blood donors. Animal studies and preliminary clinical studies suggest that C1-Inh also may be used for the treatment of various other diseases. In this chapter, we will briefly review the biochemistry and biology of C1-Inh. In addition, we will describe the methods used to evaluate the antigenic and functional state of C1-Inh.

1.1. Biochemistry of C1-Inh

C1-Inh belongs to the superfamily of serine proteinase inhibitors (serpins) *(1–4)*, of which α1-antitrypsin is considered to be archetype. C1-Inh is a major inhibitor of factor XIIa, kallikrein, and factor XIa of the contact system and the only known inhibitor of activated C1s and C1r from the classical pathway of complement *(4–11)*. Hence, C1-Inh is an important regulator of inflammatory reactions, and to a lesser extent of (factor XI-dependent) activation of clotting. The molecular weight of C1-Inh is approximately 105 kD

From: *Methods in Molecular Biology, vol. 150: Complement Methods and Protocols*
Edited by: B. P. Morgan © Humana Press Inc., Totowa, NJ

and its plasma concentration around 270 mg/L, or 1 U per mL *(4,12)*. Full-length cDNA coding for C1-Inh has been cloned *(13)*, and definitely confirmed that C1-Inh belongs to the serpin family. The part of serpins that directly interacts with target-proteinases is known as the reactive center, and is located on a protruded loop. This site functions as a bait for the proteinases. The so-called P1-P1' residues of this reactive centre loop match the specificity of the target proteinase. C1-Inh contains the amino acid residues arginine and threonine at these positions, respectively *(13)*, marking it is an inhibitor of trypsin-like serine proteinases. This agrees well with its function as an inhibitor of the major plasma cascade systems because proteinases belonging to these systems are mainly from the trypsin-branch of the serine proteinase family and prefer arginine at the P1-position of their substrates. Functional C1-Inh has been expressed in COS-cells *(14)* and recombinant C1-Inh P1 mutants expressed in this system, have confirmed the absolute requirement for arginine at the P1-position for its specificity *(15)*, although other residues such as that at the P2 position, also contribute to this specificity *(16)*. Based on the crystallographic structure of other serpins such as α1-antitrypsin and the non-inhibitory serpin ovalbumin, a three-dimensional model for C1-Inh has been proposed *(17)*.

The location of various functional sites on the C1-Inh molecule is not known although one site (Gln452-Phe455) involved in the interaction with C1s has been proposed *(18)*. Studies with truncated recombinant C1-Inh have shown that the amino-terminal domain consisting of about 100 amino acid residues, is not essential for function *(19)*. Like other serpins, C1-Inh inhibits proteinases by binding to their active site via its reactive center *(6,9,10)*. This reaction can be described according to the following equation:

$$(1) \qquad\qquad (2) \qquad\qquad (3)$$

$$\text{P + C1-inh} \longleftrightarrow \text{P.C1-Inh} \longleftrightarrow\!\!\!\twoheadrightarrow \text{P.C1-Inh*} \rightarrow \text{P + C1-Inh*}$$

$$\text{P = target-protease}$$

$$\text{C1-Inh* = modified C1-Inh}$$

The first interaction (**step 1**) yields a reversible complex between the target-protease and C1-Inh, such complexes may dissociate to some extent into an active protease and an active inhibitor. C1-Inh then undergoes conformational changes because of insertion of part of the reactive center loop, in particular the amino acid residues P14 to P10, into a 5-stranded β-sheet, to yield modified C1-Inh (C1-Inh* in the equation), which is tightly bound to the target-proteinase (**step 2**). Although this formation of complexes between a target-proteinase and modified C1-Inh has been considered to be irreversible, analysis of so-called progress curves of this reaction, indicates that this complex formation is reversible *(15)*. However, because of the very low K_{off} relative to the K_{on} of this reaction, in practice **step 2** of the equation should be considered as pseudoirreversible. The importance of the residues at P14, P12, and P10 for the conformational changes and for the inhibitory function, is underscored by the notion that these are highly conserved in inhibitory serpins, but not in noninhibitory serpins such as ovalbumin. Moreover, mutations of these residues, as occur in some patients with type II HAE (characterized by the presence of a dysfunctional C1-Inh protein), turn the inhibitor into a substrate for its target-proteinases (P14; P12) or induce a conformational change mimicking that of complexed C1-Inh (P10) *(17,20,21)*. Analysis of dysfunctional proteins resulting from other mutations has revealed that other parts of the molecule are also important. Though very stable, the complexes between modified C1-Inh and target-proteinases, dissociate very slowly into an inactive modified C1-Inh (with changed conformation because of a completed insertion of the cleaved active site loop into the 5-stranded β-sheet) and an active target-proteinase (**step 3** in the equation). The complexes formed are removed from the circulation with an apparent half-life time of clearance ranging from 20 to 47 min *(22–25)*, via receptors specific for complexed serpins, most likely the lipoprotein receptor-related protein *(26)*, on hepatocytes *(27,28)*. This rapid clearance causes most complexes between C1-Inh and target-proteinases to be cleared before they dissociate, and hence ensures efficient removal of biologically active proteinases.

The function of C1-Inh can be markedly enhanced by glucosaminoglycans (GAG) such as heparin and the semisynthetic dextran sulphate *(29,30)*. The binding site for GAG on C1-Inh is not known. Remarkably, the inactivation of C1s, C1r, and clotting factor XIa is enhanced, whereas that of kallikrein and factor XIIa is unaffected *(29,30)*.

1.2. Biology of C1-Inh

C1-Inh is an acute phase protein, plasma levels of which may increase up to twofold during uncomplicated infections *(31)*. Woo et al. *(32)* reported that the synthetic rate of C1-Inh may increase up to 2.5 times the normal rate in patients with rheumatoid arthritis. Among the cytokines that stimulate the synthesis of C1-Inh is interferon-gamma *(33)*.

C1-Inh, like most other serpins, can be inactivated by elastase released from activated neutrophils to yield modified C1-Inh *(34–36)*. Notably, this modified C1-Inh is not precisely the same as that depicted in the equation above. The latter is cleaved at the peptidyl bond between the P1 and the P1'-residues, whereas elastase-modified C1-Inh is cleaved at the bond between P4 and P5, or that between P2 and P3 *(37)*. Inactivation of C1-Inh by elastase may occur locally in inflamed tissues and contribute to local complement activation.

Modified C1-Inh circulates at detectable levels in healthy individuals, and at higher levels in patients with sepsis *(12)*. C1-Inh mutants with a decreased susceptibility for inactivation by elastase have been developed *(37)*, but their therapeutic efficacy remains to be established.

In normal volunteers, the fractional catabolic rate (FCR) of C1-Inh is 2.5% of the plasma pool per hour yielding an apparent plasma half-life time of clearance of about 28 h *(32,38)*. The half-life time of clearance of human C1-Inh in rabbits is comparable, i.e., 26 h *(39)*, whereas that in rats is considerably shorter, i.e., about 4.5 h *(22)*. The apparent half-life time of clearance has been reported to be considerably longer in patients with HAE, in whom it may be over 48 h *(40,41)*. Notably, clearance of C1-Inh in these patients is often determined by assessing the course of plasma levels following the intravenous administration of exogenous C1-Inh. This assessment may not be correct, because at lower plasma levels of C1-Inh (as occurs in untreated HAE patients) the first component of complement, C1, is autoactivated *(42)*, and consumes functional C1-Inh. At higher concentrations of C1-Inh (as occur after administration of C1-Inh) this autoactivation is inhibited leading to a decreased consumption of C1-Inh. Hence, following a therapeutic dose of C1-Inh, plasma concentrations of C1-Inh increase because of the administration of exogenous C1-Inh as well as because of a reduced consumption of endogenous C1-Inh.

C1-Inh is heavily glycosylated, the protein portion of the molecule constituting only 51% of its molecular mass *(13)*. It contains some 20 carbohydrate groups, i.e., six glucosamine-based, five galactosamine-based, remainder being linked to threonine-residues. Most carbohydrate groups are located at the N-terminal region *(13)*. Their function is largely unknown. Removal of sialic acids enormously enhances the clearance of C1-Inh from the circulation yielding an apparent half-life time of 3–5 min *(39)*, presumably via binding to asialoglycoprotein receptors in the liver. The enhanced clearance of asialo-C1-Inh is because of exposure of penultimate galactosyl residues, as the subsequent removal of the latter prolongs the clearance rate up to a value similar to that of normal C1-Inh *(39)*. Removal of sialic acid or galactose groups does not impair the functional activity of C1-Inh in vitro.

C1 is a macromolecular protein complex consisting of 1 C1q molecule and a Ca^{2+}-dependent tetramer of 2 C1r and 2 C1s molecules *(5,43)*. C1r and C1s are homologous serine proteinases, which in the tetramer are inactive zymogens. Upon binding of C1 to an activator, autoactivation of C1r is triggered, which subsequently results in activation of C1s *(44)*. In nonbound, nonactivated C1, the C1r–C1s tetramer is arranged between the stalks of C1q in such a way that the 2 C1s molecules prevent a close contact between the catalytic sites of the C1r molecules, and hence prevent autoactivation of pro-C1r *(43)*. Upon binding to an activator the conformation of C1q changes resulting in a spatial arrangement of the $C1r_2C1s_2$ that allows the catalytic sites of pro-$C1r$ to become in close contact thereby facilitating autoactivation $C1r$ and subsequent activation of $C1s$ *(43)*. In the absence of an activator, $C1$ is also autoactivated although at a slower rate than in $C1$ bound to an activator. At physiological concentrations, $C1$-Inh (which is at about sevenfold excess) by interacting with the catalytic sites of nonactivated $C1r$, prevents autoactivation of $C1$ to some extent *(42)*. During activation both $C1r$ and $C1s$ are cleaved to become two-chain proteins. Under physiological conditions the active molecules are inactivated within seconds by $C1$-Inh, resulting in the release of stable complexes consisting of $C1rC1s(C1$-Inh$)_2$ *(45)*. These complexes can be measured in plasma and reflect activation of $C1$ *(46)*.

In this chapter, we will review the biochemistry and biology of $C1$-Inh, as well as the assays for the various forms of $C1$-Inh in plasma. Assays for functional $C1$-Inh, $C1$-Inh complexed to a proteinase, as well as for inactivated $C1$-Inh are available. An analysis of $C1$-Inh in plasma is particularly helpful for the diagnosis and treatment of acquired and hereditary angioedema. An increasing number of studies suggest that $C1$-Inh therapy may be used for the treatment of various other diseases, such as sepsis, myocardial infarction, and capillary leak syndrome, as well. Therefore, the issues discussed in this chapter may be helpful for the understanding of the pathogenesis of these diseases as well.

2. Materials

2.1. Buffers and Reagents

1. Phosphate buffered saline (PBS): supplied in tablet form by Oxoid Ltd.
2. Wash buffer: PBS containing 0.1% (w/v) Tween 20.
3. Coating buffer: 0.1 *M* Sodium-bicarbonate, pH 9.0.

2.2. Chromogenic Assay for C1-Inh Activity

1. Pooled serum as a standard.
2. Microtiter plate (round bottom for preparation of dilutions; flat-bottom for actual assay).
3. A spectrophotometer capable of measuring at a wavelength of 405 nm and at various time intervals.
4. Plastic enzyme-linked immunosorbent assay (ELISA) plates (Flow Laboratories).
5. Purified $C1s$ (*see* Chapter 2). Purified active $C1s$ is also commercially available (Sigma).
6. Chromogenic substrate S-3214.
7. Complete kits for functional $C1$-Inh based on $C1s$ and chromogenic substrates are available (for example, the Berichrom C1-inactivator kit (Behringwerke AG, Marburg, Germany). This kit contains chromogenic substrate for $C1s$, substrate buffer, freeze-dried purified $C1s$.

3. Methods
3.1. Functional C1-Inh

The equation describing the interaction of C1-Inh with a target-proteinase:

$$P + C1\text{-inh} \longleftrightarrow P.C1\text{-Inh} \longleftrightarrow P.C1\text{-Inh}^* \rightarrow P + C1\text{-Inh}^*$$

where P = target-protease and C1-Inh* = modified C1-Inh indicates that C1-Inh may exist in various forms in plasma, i.e., functional C1-Inh, C1-Inh complexed to proteinases, and proteolytically inactivated C1-Inh. These various forms of C1-Inh can be measured with monoclonal antibodies recognizing epitopes specifically expressed on complexed or inactivated C1-Inh (25,47).

Functional C1-Inh in plasma is usually measured with C1s and a chromogenic substrate (for example, S-3214). In these so-called chromogenic assays, an excess of C1s (over the amount of C1-inh present in a plasma sample) is added, and after certain incubation residual amount of C1s is measured by assessing the rate of conversion of added chromogenic substrate with a spectrophotometer. These assays are very convenient and do not need expensive equipment. Results of the chromogenic assay may be expressed as a percentage of the level present in pooled plasma obtained from normal donors.

3.1.1. Chromogenic Assay for the Measurement of Functional C1-Inh

1. Appropriate dilutions of standard and samples (usually in the range of 1:10 to 1:50) are prepared in wash buffer. These are incubated for about 10–15 min at 37°C with appropriate C1s solution (for example, 25 µL sample dilution with 100 µL C1s solution). During this step C1-Inh is bound to C1s. The plate should be sealed with plastic to prevent evaporation.
2. Chromogenic substrate is added (for example, 50 µL diluted S-3214). Plates are sealed again, whereafter the conversion of the substrate with generation of a yellow color is measured using a spectrophotometer at a wave length of 405 nm at various time intervals.
3. The slope of the curve obtained indicates the amount of active C1s left in the mixtures (the slower the conversion of the substrate, the higher the level of functional C1-Inh in the sample tested).
4. The results obtained can be used to calculate the amount of C1-Inh in the sample by comparison with results obtained with pooled serum, in which the concentration of C1-Inh can be arbitrarily set at 100% or 1 U per mL (see Notes 1–5).

3.2. Antigenic C1-Inh

As explained above, various forms of C1-Inh may exist in plasma, including functional C1-Inh, C1-Inh complexed to a proteinase, and proteolytically cleaved C1-Inh. Some patients with HAE have a point mutation in the C1-Inh gene leading to the production of dysfunctional C1-Inh; these patients have

HAE type 2, which is characterized by low levels of functional C1-Inh, but normal levels of antigenic C1-Inh. In plasma from these patients additional forms of C1-Inh may circulate. These forms may be multimeric C1-Inh, or C1-Inh with changed conformation, that is noncomplexed, noncleaved C1-Inh having a conformation resembling that of complexed or cleaved C1-Inh. These forms of C1-Inh can be detected via their increased thermostability or reactivity with particular monoclonal antibodies (MAbs).

Total antigenic levels of C1-Inh can be measured with antibodies against epitopes expressed on all forms of C1-Inh. In general, most polyclonal antisera contain such antibodies. In addition, MAbs against these epitopes are also available, for example, the MAb RII *(47)*. With these antibodies regular sandwich-type ELISAs (*see* Chapter 13) can be done to determine total antigenic levels of C1-Inh.

3.3. Complexed C1-Inh

Complexes of C1-Inh with its target proteases are often used to detect activation of the proteinase in plasma. For example, complexes between C1s and C1-Inh are assessed to measure C1 activation in plasma *(46)*. These complexes are detected with so-called differential antibody assays: antibodies against one moiety of the complex are immobilized onto a solid phase, such as an ELISA plate, and incubated with samples to be tested. Bound complexes are then detected with labeled antibodies against the other moiety of the complex (*see* Chapter 7 for details of ELISA assays for complexes). In the first generation of these types of assays, antibodies recognizing all forms of the inhibitor or all forms of the proteinase were used *(46,50)*. Consequently, the immobilized antibody bound not only the complexes, but also the native inhibitor or the nonactivated proteinase. This saturation of the immobilized antibody with nonactivated components limited the sensitivity of the assay.

Therefore, a better approach is to use antibodies against neo-epitopes expressed by one of the moieties of the complex (Chapter 7). In the case of C1-Inh, complex-specific MAb indeed are available, for example, the MAb KOK12 *(25)*. Notably, the precise location of epitope for MAb KOK12 on the C1-Inh molecule is not known. The use of this MAb in assays for C1-Inh complexes enhances at least 20-fold the sensitivity of assays for C1-Inh complexes *(25)*.

3.4. Proteolytically Inactivated C1-Inh

A MAb specific for proteolytically inactivated C1-Inh, i.e., C1-Inh of which the reactive center has been cleaved at the bond between the P1 and P1'-residues or at other peptidyl bonds, has been described. This MAb, KII, has been used to determine circulating levels of inactivated C1-Inh in plasma in septic patients as well as in septic baboons *(12,51)*. Increased levels of inactivated C1-Inh have also been found in some patients with HAE and in patients with AAE *(17,52,53)*.

3.5. Autoantibodies Against C1-Inh

(Auto-)antibodies against C1-Inh may occur sometimes in patients (Chapter 13). Very infrequently, these antibodies are found in patients with HAE and may have been induced by exogenous C1-Inh prepared from pooled plasma from normal donors. As they are heterozygous deficient and hence have some endogenous C1-Inh as well, HAE patients will only rarely develop such antibodies. In our experience, we have seen only two patients with proven HAE with circulating polyclonal autoantibodies against C1-Inh. One of these also suffered from systemic lupus erythematosus (SLE), which may explain the occurrence of the autoantibodies. The other HAE patient did not have any accompanying disease that could explain the occurrence of anti-C1-Inh. In this patient, the administration of exogenous C1-Inh may have triggered the formation of anti-C1-Inh antibodies.

Autoantibodies against C1-Inh occur more frequently in patients with AAE *(54,55)*. In these patients, the antibodies are often monoclonal and point to the presence of an underlying malignant B-cell population. By binding to C1-Inh the autoantibodies turn the inhibitor into a substrate for its target proteinase, which yields increased concentrations of proteolytically cleaved C1-Inh (see previous section). For this reason, exogenous C1-Inh administered to these patients is rapidly inactivated, which explains the relative inefficiency of exogenous C1-Inh as a treatment for these patients.

Assays for (auto-)antibodies against C1-Inh can be performed with regular ELISA systems. Purified C1-Inh (most C1-Inh preparations for clinical use are >95% pure, and can be used for this purpose) is coated onto microtiter plates, and incubated with samples to be tested. Bound autoantibodies are detected by a subsequent incubation with labeled anti-IgG/M/A reagents. Anti-κ or anti-λ light chain labeled antibodies are useful to identify the monoclonal origin of the autoantibodies.

1. Microtiter plates are coated with purified C1-Inh by incubation overnight at room temperature at 2 μg/mL in coating buffer (100 μL/well).
2. Plates are then washed three times with wash buffer, blocked with 5% bovine serum albumin in wash buffer, and washed again in wash buffer.
3. Appropriate serum sample dilutions wash buffer are added to the wells, which are then incubated for 1–2 h at 37°. Pooled plasma/serum from normal donors is used as negative control.
4. The plates are washed again three times in phosphate-buffered saline (PBS) and incubated for 1 h at 37°C with peroxidase-labeled rabbit anti-human Ig (Dako), diluted 1:500 in wash buffer containing 1% rabbit serum (100 μL/well).

5. The plates are washed three times in wash buffer, once in PBS alone and developed using OPD reagent (Dako) (*see* **Notes 6** and **7**).

4. Notes

1. Results of untreated HAE patients will yield levels lower than 25% of normal. In our experience, the variation of this type of assay is around 10%, the sensitivity is at about 5% of the normal plasma level and various pitfalls may occur. The chromogenic substrate is not completely specific for C1s, but can be converted by other serine-proteinases, in particular, by those belonging to the trypsin-branch. Depending on the precise assay conditions plasma samples may nonspecifically convert the chromogenic substrate because of the presence of ill-defined trypsin-like proteinases, yielding artefactually low levels of functional C1-Inh.

2. The assay described is based on the interaction of C1-Inh with its target-protease C1s. Similar types of assays can be done using other target proteases for C1-Inh. A method based on the inhibition of kallikrein has been advised because chromogenic substrates for kallikrein are readily available. However, kallikrein is not only inactivated by C1-Inh, but also by α2-macroglobulin. Hence, this latter inhibitor has to be inactivated before measurement of C1-Inh, for example, by pre-incubation of samples with 0.2 M methylamine (final concentration), which inactivates the thioester bonds of α2-macroglobulin. Probably, for this reason, assays for functional C1-Inh based on the inactivation of kallikrein are not widely used.

3. A low functional C1-Inh level as determined with assays like that described may indicate a low level of functional C1-Inh in vivo, or an in vitro artefact. It is known that up to 20% of human serum samples stored at 0°C undergo so-called "cold-activation." This phenomenon is caused by in vitro activation of the clotting system (via factors XII and VII), and induces inactivation and consumption of functional C1-Inh yielding levels of <20% of the normal level. To discriminate between "cold-activation" and hereditary angioedema, it is advised to measure C4 levels because these are normal in case of "cold-activation," and low in case of untreated angioedema. A low level of functional C1-Inh in the presence of normal C4 level points to an in vitro artefact, at least in patients that have not received danazol or other androgenic steroids. "Cold-activation" of samples is reduced by not storing samples at around 0°C (i.e., at <–20°C or at room temperature) and by using plasma (EDTA) rather then serum.

4. Low levels of functional C1-Inh (<20% of normal) in the presence of low C4 and normal C3 levels strongly suggests C1-Inh deficiency. Antigenic levels of C1-Inh may be normal in case of the presence of dysfunctional C1-Inh (type 2 deficiency) or correspondingly low (type 1 deficiency). To discriminate between acquired and hereditary forms of C1-Inh deficiency the determination of C1q may be helpful (low in case of acquired and normal in case of hereditary C1-Inh deficiency). However, to be certain, analysis of genomic DNA and identification of the underlying defect may be warranted, especially in cases of *de novo* mutations, which have a negative family history (up to 20% of C1-Inh hereditary deficiencies are *de novo* mutations).

5. Another point that merits attention, is that the assay performed as described above, is an end-stage assay in that it measures the total amount of functional C1-Inh in a sample. The inhibitory function of C1-Inh may be enhanced by several GAG such as heparin *(29,48)*. These GAGs do not change the molar ratio of C1s or C1-Inh in a given sample, but allow a much more rapid complex formation between the proteinase and the inhibitor. Thus, in the presence of GAG C1-Inh more rapidly inactivates C1s, and consequently the half-life time of active C1s is shortened. In an end-stage assay, such an enhanced inhibition is not measured. Hence, this assay may underestimate the inhibitory function of C1-Inh. Various alternative assays for functional C1-Inh have been described including esterolytic or hemolytic assays. As the chromogenic assay is the easiest one to perform, these assays are less frequently used. A particular type of immunoassay can also be used to assess functional C1-Inh in a sample. In this assay, immobilized purified C1s is allowed to interact with C1-Inh in a sample. As with fluid-phase C1s, functional C1-Inh will form stable complexes with immobilized C1s, which is detected by a subsequent incubation with anti-C1-Inh antibodies. This type of assay correlates well with a chromogenic assay *(49)*, and because of its high sensitivity is very useful when low concentrations of C1-Inh are tested. An alternative approach is to allow C1-Inh in plasma to bind to an immobilized antibody. Bound functional C1-Inh is then specifically detected by incubation with biotinylated C1s *(22)*.

6. Autoantibodies in our view should be further investigated to determine their polyclonal or monoclonal nature (using anti-kappa or antilambda light chain antibodies in the detection **step 4**). Autoantibodies should also be assessed for their functional activity on C1-Inh. This can be done by mixing patient plasma containing anti-C1-Inh antibodies with normal plasma and measuring residual functional C1-Inh (see above). Alternatively, purified C1-Inh can be incubated with affinity purified patient IgG (purified by protein G chromatography), whereafter the function of C1-Inh is assessed.

5. Autoantibodies against C1-Inh are the cause of many cases of acquired C1-Inh deficiency. They may be formed by a malignant B-cell clone, and in that case are monoclonal sometimes manifesting themselves as paraprotein on native gel-electrophoresis. Hence, a diagnosis of acquired C1-Inh deficiency should always lead to a search for the presence of a B-cell tumor.

References

1. Travis, J. and Salvesen, G. S. (1983) Human plasma proteinase inhibitors. *Annu. Rev. Biochem.* **52,** 655–709.
2. Carrell, R. W. and Boswell, D. R. (1990) Serpins: the superfamily of plasma serine proteinase inhibitors, in *Proteinase Inhibitors* (Barrett, A. J. and Salvesen, G., eds.), pp. 403–420.
3. Carrell, R. W. and Travis, J. (1985) 1-antitrypsin and the serpins: variation and countervariation. *Trends Biochem. Sci.* 20–24.
4. Schapira, M., Agostini de, A., Schifferli, J. A., and Colman, R. W. (1985) Biochemistry and pathophysiology of human C1 inhibitor: Current issues. *Complement* **2,** 111–126.

5. Cooper, N. R. (1985) The classical complement pathway: activation and regulation of the first complement component. *Adv. Immunol.* **37,** 151–216.

6. Sim, R. B., Reboul, A., Arlaud, G. J., Villiers, C. L., and Colomb, M. G. (1979) Interaction of ^{125}I-labelled complement subcomponents C1r and C1s with protease inhibitors in plasma. *FEBS Lett.* **97,** 111–115.

7. Chan, J. Y. C., Burrowes, C. E., Habal, F. J., and Movat, H. Z. (1977) The inhibition of activated factor XII (Hageman factor) by antithrombin III: effect of other plasma proteinase inhibitors. *Biochem. Biophys. Res. Commun.* **74,** 150–158.

8. Schapira, M., Scott, C. F., and Colman, R. W. (1982) Contribution of plasma protease inhibitors to the inactivation of kallikrein in plasma. *J. Clin. Invest.* **69,** 462–468.

9. van der Graaf, F., Koedam, J. A., and Bouma, B. N. (1983) Inactivation of kallikrein in human plasma. *J. Clin. Invest.* **71,** 149–158.

10. Pixley, R. A., Schapira, M., and Colman, R. W. (1985) The regulation of human factor XIIa by plasma proteinase inhibitors. *J. Biol. Chem.* **260,** 1723–1729.

11. Wuillemin, W. A., Minnema, M. C., Meijers, J. C. M., Roem, D., Eerenberg, A. J. M., Nuijens, J. H., ten Cate, H., and Hack, C. E. (1995) Inactivation of factor XIa in human plasma assessed by measuring factor XIa-protease inhibitor complexes: major role for C1-inhibitor. *Blood* **85,** 1517–1526.

12. Nuijens, J. H., Eerenberg-Belmer, A. J. M., Huijbregts, C. C. M., Schreuder, W. O., Felt-Bersma, R. J. F., Abbink, J. J., Thijs, L. G., and Hack, C. E. (1989) Proteolytic inactivation of plasma C1-Inhibitor in sepsis. *J. Clin. Invest.* **84,** 443–450.

13. Bock, S. C., Skriver, K., Nielsen, E., Thogersen, H. C., Wiman, B., Donaldson, V. H., et al. (1986) Human C1 inhibitor: primary structure, cDNA cloning, and chromosomal localization. *Biochem.* **25,** 4292–4301.

14. Eldering, E., Nuijens, J. H., and Hack, C. E. (1988) Expression of functional human C1 inhibitor in COS cells. *J. Biol. Chem.* **263,** 11776–11779.

15. Eldering, E., Huijbregts, C. C. M., Lubbers, Y. T. P., Longstaff, C., and Hack, C. E. (1992) Characterization of recombinant C1 inhibitor P1 variants. *J. Biol. Chem.* **267,** 7013–7020.

16. Zahedi, R., Wisnieski, J., and Davis, A. E., III (1997) Role of the, P. 2., residue of complement 1 inhibitor (ala443) in determination of target protease specificity. Inhibition of complement and contact proteases. *J. Immunol.* **159,** 983–988.

17. Skriver, K., Wikoff, W. R., Patston, P. A., Tausk, F., Schapira, M., Kaplan, A. P., and Bock, S. C. (1991) Substrate properties of C1 inhibitor Ma (alanine->glutamic acid). Genetic and structural evidence suggesting that the P12-region contains critical determinants of serine protease inhibitor/substrate status. *J. Biol. Chem.* **266,** 9216–9221.

18. He, S., Yang J-C, Tsang, S., Sim, R. B., and Whaley, K. (1997) Role of the distal hinge of C1-inhibitor in the regulation of C1s activity. *FEBS Lett.* **412,** 506–510.

19. Coutinho, M., Aulak, K. S., and Davis, A. E., III (1994) Functional analysis of the serpin domain of C1-inhibitor. *J. Immunol.* **153,** 3648–3654.

20. Aulak, K. S., Eldering, E., Hack, C. E., Lubbers, Y. P., Harrison, R. A., Mast, A., Cicardi, M., and Davis, A. E., III (1993) A hinge region mutation in C1-inhibitor

(Ala436->Thr) results in nonsubstrate-like behavior and in polymerization of the molecule. *J. Biol. Chem.* **268**, 18,088–18,094.

21. Davis, A. E., III, Aulak, K. S., Parad, R. B., Stecklein, H. P., Eldering, E., Hack, C. E., et al. (1992) C1 inhibitor hinge region mutations produce dysfunction by different mechanisms. *Nat Genet.* **1**, 354–358.

22. de Smet, B. J. G. L., De Boer, J. P., Agterberg, J., Rigter, G., Bleeker, W. K., and Hack, C. E. (1993) Clearance of human native, proteinase-complexed, and proteolytically inactivated C1-inhibitor in rats. *Blood* **81**, 56–61.

23. Wuillemin, W. A., Bleeker, W. K., Agterberg, J., Rigter, G., ten Cate, H., and Hack, C. E. (1996) Clearance of human factor XIa-inhibitor complexes in rats. *Br. J. Haematol.* **93**, 950–954.

24. Wuillemin, W. A., Hack, C. E., Bleeker, W. K., Biemond, B. J., Levi, M., and ten Cate, H. (1996) Inactivation of factor XIa in vivo: Studies in chimpanzees and in humans. *Thromb. Haemost.* **76**, 549–555.

25. Nuijens, J. H., Huijbregts, C. C. M., Eerenberg, A. J. M., Abbink, J. J., Strack van Schijndel, R. J. M., Felt-Bersma, R. J. F., Thijs, L. G., and Hack, C. E. (1988) Quantification of plasma factor XIIa-C1-Inhibitor and kallikrein-C1-Inhibitor complexes in sepsis. *Blood* **72**, 1841–1848.

26. Storm, D., Herz, J., Trinder, P., and Loos, M. (1997) C1 inhibitor-C1s complexes are internalized and degraded by the low density lipoprotein receptor-related protein. *J. Biol. Chem.* **272**, 31043–31050.

27. Perlmutter, D. H., Glover, G. I., Rivetna, M., Schasteen, C. S., and Fallon, R. J. (1990) Identification of a serpin-enzyme complex receptor on human hepatoma cells and human monocytes. *Proc. Natl. Acad. Sci. USA* **87**, 3753–3757.

28. Pizzo, S. V., Mast, A. E., Feldman, S. R., and Salvesen, G. (1988) In vivo catabolism of 1-antichymotrypsin is mediated by the serpin receptor which binds 1-proteinase inhibitor, antithrombin III and heparin cofactor II. *Biochim. Biophys. Acta* **967**, 158–162.

29. Wuillemin, W. A., T. E., Velthuis, H., Lubbers, Y. T. P., De Ruig, C. P., Eldering, E., and Hack, C. E. (1997) Potentiation of C1 inhibitor by glycosaminoglycans. Dextran sulfate species are effective inhibitors of in vitro complement activation in plasma. *J. Immunol.* **159**, 1953–1960.

30. Wuillemin, W. A., Eldering, E., Citarella, F., De Ruig, C. P., Ten Cate, H., and Hack, C. E. (1996) Modulation of contact system proteases by glycosaminoglycans. Selective enhancement of the inhibition of factor XIa. *J. Biol. Chem.* **271**, 12,913–12,918.

31. Kalter, E. S., Daha, M. R., Ten Cate, J. W., Verhoef, J., and Bouma, B. N. (1985) Activation and inhibition of Hageman factor-dependent pathways and the complement system in uncomplicated bacteremia or bacterial shock. *J. Infect. Dis.* **151**, 1019–1027.

32. Woo, P., Lachmann, P. J. L., Harrison, R. A., and Amos, N. (1985) Simultaneous turnover of normal and dysfunctional C1 Inhibitor as a probe of in vivo activation of C1 and contact activatable proteases. *Clin. Exp. Immunol.* **61**, 1–8.

33. Heda, G. D., Mardente, S., and Schmaier, A. H. (1990) Interferon gamma increases in vitro and in vivo expression of C1 inhibitor. *Blood* **75**, 2401–2407.

34. Carrell, R. W., Christey, P. B., and Boswell, D. R. (1987) Serpins: antithrombin and other inhibitors of coagulation and fibrinolysis. Evidence from amino acid sequences, in *Thrombosis and Haemostasis* (Verstraete, M., Vermylen, J., Lijnen, R., and Arnout, J., eds.), Leuven University Press, Leuven, pp. 1–15.

35. Weiss, S. J. (1989) Tissue destruction by neutrophils. *New England J. Med.* **320**, 365–376.

36. Brower, M. S. and Harpel, P. C. (1982) Proteolytic cleavage and inactivation of 2-plasmin inhibitor and C1-inactivator by human polymorphonuclear leucocyte elastase. *J. Biol. Chem.* **257**, 9849–9854.

37. Eldering, E., Huijbregts, C. M., Nuijens, J. H., Verhoeven, A. J., Hack, C. E. (1993) Recombinant C1 inhibitor P5/P3 variants display resistance to catalytic inactivation by stimulated neutrophils. *J. Clin. Invest.* **91**, 1035–1043.

38. Quastel, M., Harrison, R., Cicardi, M., Alper, C. A., and Rosen, F. S. (1983) Behavior in vivo of normal and dysfunctional C1 inhibitor in normal subjects and patients with hereditary angioneurotic edema. *J. Clin. Invest.* **71**, 1041–1046.

39. Minta, J. O. (1981) The role of sialic acid in the functional activity and the hepatic clearance of C1-Inh. *J. Immunol.* **126**, 245–249.

40. Agostoni, A., Bergamaschini, L., Martignoni, G., Cicardi, M., and Marasini, B. (1980) Treatment of acute attacks of hereditary angioedema with C1-inhibitor concentrate. *Ann. Allergy* **44**, 299–301.

41. Gadek, J. E., Hasea, S. W., Gelfand, J. A., Santaella, M., Wickerhauser, M., Triantaphyllopoulos, D. C., and Frank, M. M. (1980) Replacement therapy in hereditary angioedema. Successful treatment of acute episodes of angioedema with partly purified C1 inhibitor. *New England J. Med.* **302**, 542–546.

42. Ziccardi, R. J. (1982) A new role for C1-inhibitor in homeostasis: control of activation of the first component of human complement. *J. Immunol.* **128**, 2505–2508.

43. Schumaker, V. N., Zavodsky, P., and Poon, P. H. (1987) Activation of the first component of complement. *Annu. Rev. Immunol.* **5**, 21–42.

44. Ziccardi, R. J. (1982) Spontaneous activation of the first component of human complement (C1) by an intramolecular autocatalytic mechanism. *J. Immunol.* **128**, 2500–2504.

45. Ziccardi, R. J. and Cooper, N. R. (1979) Active disassembly of the first complement component C1 by C1-inactivator. *J. Immunol.* **123**, 788–792.

46. Hack, C. E., Hannema, A. J., Eerenberg AJM, Out, T. A., and Aalberse, R. C. (1981) A C1-inhibitor-complex assay (INCA): a method to detect C1 activation in vitro and in vivo. *J. Immunol.* **127**, 1450–1453.

47. Nuijens, J. H., Huijbregts CCM, van Mierlo, G. M., Hack, C. E. (1987) Inactivation of C1 inhibitor by proteases: demonstration by a monoclonal antibody of a neodeterminant on inactivated, non-complexed C1 inhibitor. *Immunol.* **61**, 387–389.

48. Rent, R., Myhrman, R., Fiedel, A., and Gewurz, H. (1976) Potentiation of C1 esterase inhibitor activity by heparin. *Clin. Exp. Immunol.* **23**, 264–271.

49. Cugno, M., Nuijens, J. H., Hack, C. E., Eerenberg, A. J. M., Frangi, D., Agostoni, A., and Cicardi, M. (1990) Plasma levels of C1 inhibitor complexes and cleaved

C1 inhibitor in patients with hereditary angioneurotic edema. *J. Clin. Invest.* **85,** 1215–1220.

50. Nuijens, J. H., Huijbregts CCM, Cohen, M., Navis, G. O., de Vries, A., Eerenberg, A. J. M., Bakker, J. C., and Hack, C. E. (1987) Detection of activation of the contact system of coagulation in vitro and in vivo: quantitation of activated Hageman factor-C1-Inhibitor and kallikrein-C1-Inhibitor complexes by specific radioimmunoassays. *Thromb. Haemost.* **58,** 778–785.

51. De Boer, J. P., Creasey, A. A., Chang, A., Roem, D., Eerenberg, A. J. M., Hack, C. E., Taylor, F. B., Jr. (1993) Activation of the complement system in baboons challenged with live *E. coli:* Correlation with mortality and evidence for a biphasic activation pattern. *Infect. Immun.* **61,** 4293–4301.

52. Jackson, J., Sim, R. B., Whaley, K., and Feighery, C. (1989) Autoantibody facilitated cleavage of C1-inhibitor in autoimmune angioedema. *J. Clin. Invest.* **83,** 698–707.

53. Cicardi, M., Beretta, A., Colombo, M., Gioffre, D., Cugno, M., and Agostoni, A. (1996) Relevance of lymphoproliferative disorders and of anti-C1 inhibitor autoantibodies in acquired angio-edema. *Clin. Exp. Immunol.* **106,** 475–480.

54. Jackson, J., Sim, R. B., Whelan, A., and Feighery, C. (1986) An IgG autoantibody which inactivates C1-inhibitor. *Nature* **323,** 23–29.

13

Autoantibodies to Complement Components

Kevin A. Davies and Peter Norsworthy

1. Introduction

The purpose of the immune system is to defend the host from constantly changing microbial pathogens. Autoimmune diseases develop as a consequence of the production of antibodies and/or cells that react with self-antigens, and may recruit other effector mechanisms that result in tissue damage. Thus, in this context, autoimmunity represents an immune response to self-antigens that is sufficient to cause disease. It should not be forgotten that apparently harmless autoantibodies may also be formed following tissue damage (e.g., antiheart antibodies, after a myocardial infarction), and the evidence that many of the autoantibodies routinely measured in the laboratory are directly pathogenic, is weak.

The role of autoimmunity in many diseases is still unclear. However, the production of organ-specific autoantibodies clearly has a direct role in the pathogenesis of certain autoimmune disorders. Some of the best understood from a mechanistic point of view are myasthenia gravis, Graves' disease, and several variants of insulin resistance.

This chapter is specifically devoted to autoantibodies directed against complement (C) components. So far, all the anticomplement antibodies that have been described are directed against neoepitopes on activated or comformationally altered molecules that are not present in the native state. This suggests that certain individuals are not tolerized to these newly exposed regions, and following a triggering mechanism, they generate autoantibodies to activated C components. However, what makes one person generate C1 inhibitor (C1-inh) autoantibodies and another C3 nephritic factor is still unclear.

From: *Methods in Molecular Biology, vol. 150: Complement Methods and Protocols*
Edited by: B. P. Morgan © Humana Press Inc., Totowa, NJ

1.1. Autoantibodies to the Collagen-Like Region (CLR) of C1q

It has been demonstrated that the anti-C1qCLR autoantibody can only bind solid-phase C1q. This phenomenon was first explained by Antes et al., who demonstrated that anti-C1q antibodies are directed against neoepitopes that become exposed once the molecule is bound in the solid phase *(1)*. This argument was supported by previous studies with monoclonal antibodies *(2,3)*. Bearing these results in mind, it is now clear that the anti-C1qCLR antibodies interfere with the C1q solid-phase binding assay for the detection of circulating immune complexes (ICs), which is one of the reasons for the gradual decline of C1q-based assays for the measurement of ICs as a disease marker in systemic lupus erythematosus (SLE).

New assay systems have since been developed that directly test for the presence of anti-C1q antibodies. These are generally enzyme-linked immunosorbent assays (ELISA)-based systems with the CLR of C1q or intact C1q molecule immobilized on the plate. Since the development of these tests, anti-C1qCLR antibodies have been detected in many different diseases: SLE, membranous glomerulonephritis, rheumatoid vasculitis, and hypocomplementemic urticarial vasculitis *(4–11)*. IgG anti-C1q antibodies are found in around 30% of an unselected SLE population *(12)*. These autoantibodies show a strong inverse correlation with C4 and C3 levels in SLE *(1,4,12–14)*, and are strongly correlated with lupus nephritis disease activity *(10,12,14)*. In 1995, Coremans and coworkers demonstrated that of all the known SLE autoantibodies, IgG anti-C1qCLR antibodies had the highest predictive value for renal relapses in SLE *(15)*. SLE IgG C1qCLR autoantibodies are mainly restricted to the IgG_2 isotype. Our laboratory has recently shown that they are associated with the FcγRIIA-R131 allele that encodes a variant of the receptor that is unable to bind and effectively process IgG_2 containing immune complexes *(16)*.

IgG anti-C1q antibodies have an extremely high prevalence in hypocomplementemic urticarial vasculitis syndrome (HUVS). All of our patients with this disorder are positive for IgG anti-C1qCLR. This concurs with the work done by Wisnieski's group, where all 18 of their patients were positive *(17)*. In fact, in their report they favored a positive IgG anti-C1qCLR antibody test (in conjunction with other serological and clinical findings) as an absolute diagnostic criterion for the diagnosis of HUVS. Having said this, the presence of IgG anti-C1qCLR antibodies is not specific for HUVS, or any other condition. Because IgG anti-C1q antibodies have been described in various other systemic and renal disorders such as rheumatoid vasculitis *(5)*, antiglomerular basement-membrane disease *(11)*, membranoproliferative nephritis *(6)*, rheumatoid arthritis, mixed connective-tissue disease, and Felty's syndrome *(10)*, a positive test must be carefully considered in the context of all other clinical and serological findings for a correct diagnosis to be made.

1.2. Autoantibodies to C1 Inhibitor

C1 inhibitor (C1-Inh) is a highly glycosylated single-chain polypeptide (MW 106 kDa) that belongs to a family of serine protease inhibitors, collectively known as serpins, which circulate in the plasma (Chapter 12). C1-Inh inactivates the enzymes of the contact phase (activated Hageman factor), kinin (kallikrein), coagulation (factor XIa), and fibrinolytic (plasmin) pathways. Interestingly, it is the only serum serpin that is able to inactivate the proteolytic activity of C1r and C1s of the complement classical pathway *(18)*. Its absence, therefore, results in severely dysregulated activation of C1. The most common situation where this occurs is in the context of the genetically determined "hereditary angioedema," which is caused by the presence of either a null, or a defective C1-Inh allele (Chapter 12).

The two forms of acquired C1-Inh deficiency (described below) differ from hereditary angioedema in that the onset of the disease occurs later in life in individuals who have no family history to suggest the genetically determined form of the disorder. The complement profiles of these patients show that they have low levels or the complete absence of C1-Inh, CH50, C1q, C4, and C2. The low level of C1q is an important characteristic that distinguishes them from the hereditary form of the disease. The decreased levels of the complement components are believed to be caused by the uncontrolled activation of the classical pathway of complement. Acquired C1-Inh deficiency is also clinically associated with a range of lymphoproliferative disorders *(19,20)*.

Malbran and colleagues suggested two ways in which an autoantibody–C1-Inh interaction might inhibit the normal function of C1-Inh. They hypothesized that the autoantibody bound to a site at or near the reactive center of C1-Inh and this could either: (1) inhibit the tertiary rearrangement or sterically impede enzyme access to the reactive center of the inhibitor, or (2) successfully compete with the target enzyme by binding to an epitope in the cleaved inhibitor that includes or obscures the reactive center *(21)*. The hypothesis that these antibodies bound to the reactive center of the C1-Inh was further reinforced in 1994 by the discovery in a patient with acquired C1-Inh deficiency of an autoantibody that was able to bind to the reactive center of C1-Inh *(22)*. This antibody was able to block the ability of C1-Inh to inhibit the hydrolysis of N-carbobenzyloxy-L-lysine thiobenzylester by purified C1s. Several groups have subsequently used ELISAs to demonstrate the presence of C1-Inh autoantibodies in acquired C1-Inh deficiency *(20,23,24)*.

1.3. Autoantibodies to C3 Convertase (C3 Nephritic Factor/ C4 Nephritic Factor)

C3 nephritic factor (C3 NeF) was first described in patients with membranoproliferative glomerulonephritis (MPGN) *(25)* and partial lipodystrophy *(26)*.

It was thought that the presence of C3 NeF in the sera was responsible for the persistent hypocomplementemia seen in MPGN *(27,28)*. A few years later, it was found that C3 NeF is an IgG autoantibody that acts by stabilizing the C3/C5 convertase (C3bBb) of the alternative pathway. This prolongs its half-life (inhibition of intrinsic decay) and blocks the action of the control proteins, factors H and I (fH and fI, respectively) (inhibition of extrinsic decay) *(29,30)*, thereby promoting the continued activation of C3.

Since their initial discovery, several different types of C3 NeF have been described: properdin-independent *(31)*, properdin-dependent *(31)*, and non-hypocomplementemia-associated C3 NeF *(32)*. However, their relationship to the pathogenesis of MPGN and their interaction with the complement components is still not fully understood. The findings were not wholly unexpected as there is more than one epitope in C3bBb and variable binding to these epitopes could produce the different characteristics seen in patients. This hypothesis is supported by an early report by Daha and Vanes who showed that C3 NeF binds not only to Bb, but also to C3b in the C3bBb complex in some cases *(33)*.

An analog of C3 NeF has been found for the classical pathway C3 convertase. C4 NeF was first found in the serum of a patient with acute postinfectious glomerulonephritis *(34)*. It has since been found in patients with SLE *(35)*, chronic glomerulonephritis *(36)*, and hypocomplementemic MPGN *(37)*. C4 NeF, like C3 NeF, was also identified as an IgG autoantibody.

In 1985, Gigli et al. demonstrated that C4 NeF could interfere with all three mechanisms for the control of the fluid-phase C4b2a, it: (1) stabilized the labile convertase; (2) prevented the dissociation of C2a caused by C4bp; and (3) inhibited inactivation of C4b by fI *(38)*. Shortly after this, Fujita et al. showed that the C4 NeF present in their patient with chronic glomerulonephritis bound specifically to C4b2a and not to C4b or C3bBb *(36)*. From this, they concluded that C4 NeF was an IgG autoantibody directed against neoepitopes on the C4b2a complex that prevents the decay of *C2a* from the C4b2a, thus increasing the half-life of the convertase *(38)*.

Ohi and Yasugi have recently investigated the prevalence of both C3 NeF and C4 NeF in the sera of 100 hypocomplementemic MPGN patients. They found that out of the 100 patients with a C3 <40%, 31 were NeF positive. Of these, 12 were positive for C3 NeF only, 9 for C4 NeF only, and 10 were positive for both NeFs *(39)*. Based on these results and the coexistence of C3 NeF and C4 NeF in patients with MPGN, Ohi and Yasugi concluded that C4 NeF is also an important factor in the complement activation seen in these patients. However, despite all these data, the origin and pathological significance of C4 NeF, like C3 NeF, in the disease process still remains to be elucidated.

1.4. Other Autoantibodies in the Complement System

At a recent complement workshop a group disclosed the existence of an unique antifactor H autoantibody, a λ-light chain dimer ($λ_L$) that was present in the serum and urine of a hypocomplementemic MPGN patient *(40)*. This autoantibody was shown to bind to fH, and on binding was able to restrict fH interaction with both activator and nonactivator-bound C3b. $λ_L$ was also shown to inhibit the enzymatic inactivation of fluid-phase C3b. This factor was thus able to deplete the alternative pathway hemolytic activity of serum by inhibiting one of its regulators. However, these are the only details available at present as a recent literature survey revealed nothing further on this subject.

He and Lin *(41)* screened SLE patients for a number of anticomplement antibodies and found that 7 of 15 patients were positive for anti-C1s autoantibodies. This finding led them to functional studies, exploring the way in which the autoantibody may interact with its substrate. This group now suggests that C1s autoantibodies may stimulate the proteolytic activities of C1s leading to complement consumption and hence be a contributing factor to the hypocomplementemia seen in many SLE patients.

2. Materials

2.1. Buffers

1. Phosphate buffered saline (PBS), pH 7.2: 137 mM NaCl, 2.7 mM KCl, 8.1 mM Na_2HPO_4, 1.5 mM KH_2PO_4.
2. Complement fixation test diluent (CFD) (Oxoid, Basingstoke, Hants), pH 7.2: 4 mM Na barbitone, 145 mM NaCl, 0.83 mM $MgCl_2$, 0.25 mM $CaCl_2$.
3. Glycine barbitone Tris buffer, pH 7.2 (GBT): 63 mM Na barbitone, 11 mM barbitone, 187 mM glycine, 72 mM Tris HCl.
4. Veronal buffered saline (VBS), pH 7.2: 150 mM NaCl, 1.8 mM Na barbitone, and 3.2 mM barbitone.

2.2. Materials for ELISA Assays

1. Multiwell plates: Titertek™ Immunoassay plates (Flow Laboratories) or equivalent.
2. Plate sealing tape.
3. ELISA plate reader capable of reading absorbance at multiple wavelengths.
4. Carbonate/bicarbonate coating buffer, pH 9.6: Mix the required volumes of 200 mM stock solutions of Na_2CO_3 and $NaHCO_3$ to achieve the correct pH.
5. Wash buffer, PBS containing 0.075% Tween-20.
6. Blocking buffer, PBS containing 1% newborn calf serum (buffer A) or 2% bovine serum albumin (buffer B).
7. Appropriate capture and detection antibodies against the antigen of interest.
8. Appropriate enzyme-labeled tertiary antibodies—antirabbit IgG, antimouse IgG, or antihuman IgG labeled with alkaline phosphatase (AP) or horseradish peroxidase

(HRP), available from numerous sources, including Bio-Rad, Dako, Sigma, Jackson Immunoresearch, and so on.

9. Appropriate substrate for antibody-conjugated enzyme. For AP, p-nitrophenyl phosphate (pNPP) tablets (Sigma), for HRP, 1,2-phenylenediamine dihydrochloride (OPD) tablets (Dako; Sigma). Follow manufacturer's protocol to make up substrate solutions.
10. Stop solution to prevent further color development; 3 M NaOH for AP substrates and 2 M sulphuric acid for HRP substrates. Add 50 μL to each well.

2.3. Materials for SDS PAGE and Western Blotting

1. Sodium dodecyl sulfate-polyacrylamide gel electrophoresis (SDS PAGE) running buffer, pH 8.3: 5 mM Tris base, 38.4 mM glycine, 0.02% (w/v) SDS.
2. Western blotting transfer buffer pH 8.3: 39 mM glycine, 48 mM Tris base, 1.3 mM SDS, and 60% (v/v) methanol.
3. Acrylamide, *bis*-acrylamide buffering solutions and catalysts for production of gels as detailed in methods Premade solutions of acrylamide and *bis*-acrylamide are widely available commercially that are very reliable and a good deal safer to handle than acrylamide in its powdered form (*see* **Note 1**). Prepoured ready-to-use gels are now also widely available.
4. Sample loading buffer comprising 75 mM Tris/HCl pH 6.9, 1% (w/v) SDS, 10% (v/v) glycerol, and 0.1% bromophenol blue. "Reduced" sample loading buffer additionally contains 1% (w/v) β-mercaptoethanol.
5. Apparatus for running SDS-PAGE gels available in various sizes and formats from Bio-Rad, Amersham Pharmacia, and many other suppliers. Automated electrophoresis systems such as the Phastsystem developed by Pharmacia are also now available.
6. Apparatus for Western blotting of SDS-PAGE gels, again available in many formats from many suppliers.
7. Power packs for above apparatus.
8. Nitrocellulose or other appropriate blotting membrane.
9. Blocking solution: PBS containing 3% bovine serum albumin or PBS containing 4% nonfat freeze-dried milk powder.
10. Wash buffer: PBS containing 0.5% Tween-20
11. Appropriate detection antibodies and HRP-labeled secondary antibodies.
12. HRP substrate solution, 4-chloro-1-naphthol tablets (Sigma) dissolved according to the manufacturer's instructions. Sensitive HRP substrate systems based on enhanced chemiluminescence (ECL) and exposure onto X-ray film are now widely available (Amersham Pharmacia, Pierce, and others).

3. Methods
3.1. ELISA

The most widely used assay to screen for antibodies is the ELISA. This can be either a direct ELISA where the antigen is immobilized straight on to a

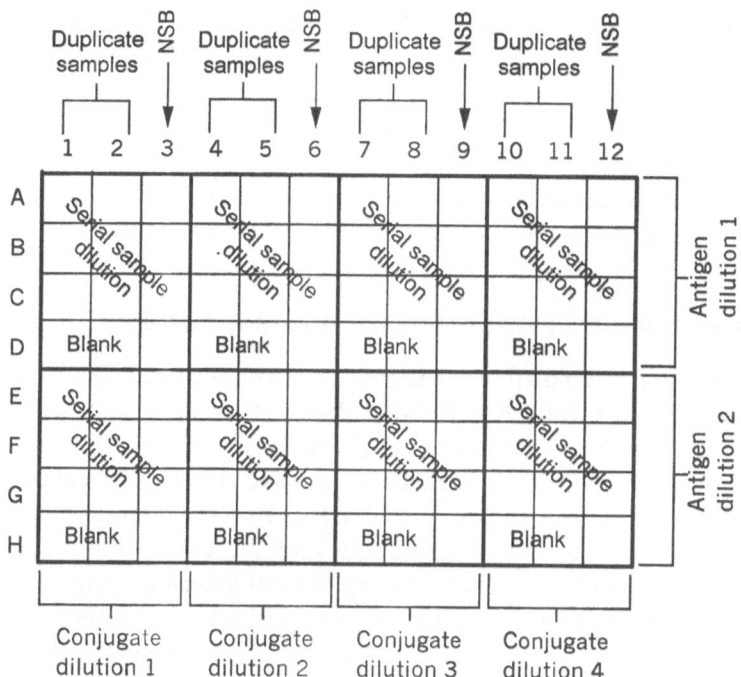

Fig. 1. A checkerboard titration for the establishment of an ELISA. Nonspecific binding wells (NSB) are indicated.

multiwell plate, or a sandwich ELISA where the antigen is captured using a plate that has been precoated with an antibody. The usefulness of ELISAs to screen for the presence of anticomplement antibodies has recently been demonstrated by He and Lin who employed this technique to screen for a range of complement autoantibodies *(41)*.

The correct concentration for each step of the ELISA (i.e., capture antibody, antigen concentration, serum dilution, and conjugate dilution) will need to be titrated out in order to create a reproducible, accurate assay. The establishment of an ELISA is usually achieved using a so-called checkerboard titration of all the variables (*see* **Fig. 1**).

When an ELISA system is first set up, it is necessary to have nonspecific binding wells. These are wells that have not been coated with the antigen, but contain all the other components. In this way, a more accurate assessment of the autoantibody binding can be made. Some ELISAs may have negligible nonspecific binding so this control may be omitted once the test has been established.

The two most commonly used conjugate systems are alkaline phosphatase and HRP. Ideally, an AP-based ELISA should develop in approx 30 min. One of the simplest ways to adjust the development time after the assay has been established is by altering the dilution of conjugate. HRP-based ELISA tend to develop faster, in approx 10–20 min. The appropriate stop solution is added at the end of this period.

If weak signals are obtained in either Western blots or ELISAs, signal can be amplified using a biotin/streptavidin system, however, this may give problems with high background as a consequence of inappropriate amplification of signal from nonspecifically bound protein.

3.2. Measurement of Anti-C1qCLR Autoantibodies

In recent years, two different ELISA systems have been specifically developed to detect anti-C1qCLR antibodies, but to prevent false positive readings arising from the nonspecific interaction between the globular heads of C1q and the Fc region of immunoglobulins. These involve either: (1) incubating the test serum in high ionic strength buffer, which modifies the interaction between the IC and the globular heads of C1q, but does not affect the high avidity binding of its autoantibody *(7)*, or (2) employing the collagen-like stalks as the solid-phase antigen (utilizing the fact that the autoantibodies are directed against the CLR of the C1q) *(4)*. There still remains some controversy over how effective high-ionic-strength buffers are at dissociating the interaction between ICs and C1q. However, in our experience there is an extremely good correlation between the high salt and the collagen-like stalk ELISAs, indicating that they are detecting the same autoantibodies.

3.2.1. Saline ELISA for Detecting Human Anti-C1q Antibodies

Human C1q autoantibodies are measured using a direct ELISA technique. This method is adapted from an assay developed by Siegert et al. *(7)*.

1. Coat 96-well flat-bottomed plates with 100 µL per well of C1q whole molecule at 2 µg/well diluted in coating buffer by incubation overnight at 4°C. Include non-specific binding wells (not coated with C1q). Include standard positive sample wells.
2. Wash plates twice with wash buffer and twice with PBS.
3. Block plates with 200 µL/well-blocking buffer A for 2 h at 37°C.
4. Wash plates as before.
5. Centrifuge all patient samples at 8000g in a benchtop microcentrifuge for 5 min before dilution to remove immune complexes.
6. Dilute serum samples 1:50 in wash buffer containing 1% newborn calf serum and 1 M NaCl, and add 100 µL/well to the plates in duplicate. Incubate for 1 h at 37°C.
7. Wash plates as before.
8. Add 100 µL/well AP-conjugated goat antihuman IgG (gamma-chain specific) antibody diluted 1:5000 in wash buffer containing 1% newborn calf serum, incubate for 1 h at 37°C.

9. Wash as before.
10. Add 100 µL/well of AP substrate solution and allow color to develop in the dark at room temperature for 10–60 min. Add stop solution (50 µL/well). Read the absorbance at 405 nm in an ELISA plate reader.
11. Subtract nonspecific binding values from specific to correct for background binding to plate.
12. The standard positive sample, included in each ELISA plate, is arbitrarily assigned a value of 100; a conversion factor (CF) can then be derived from its corrected absorbance reading at 405 nm (OD_{405nm}). The corrected OD_{405nm} reading obtained for each test sample is then multiplied by the CF. In this way, all the samples in each subsequent ELISA can be related to one another.

Conversion Factor (CF) = 100 Arbitrary Units/OD_{405nm} positive control

CF X OD_{405nm} of sample = Result in Arbitrary ELISA units (Arbitrary EU)

3.2.2. A CLR "Stalk" ELISA to Measure Anti-C1q Antibodies

This is an alternative ELISA technique to the method described above that utilizes pepsin digested C1q and was devised by Wener and colleagues in 1989 *(4)*. Pepsin digests away the globular heads of C1q that nonspecifically recognize the Fc and therefore removes the need for 1 *M* NaCl in the dilution buffer. The CLR of C1q (which contains the neoepitopes to which C1q autoantibodies are generated) remains intact.

1. Coat 96-well flat-bottomed plates overnight at 4°C with 100 µL/well of C1qCLR diluted to 10 µg/mL in PBS. Include nonspecific binding wells.
2. Wash plates twice with wash buffer and twice with PBS.
3. Block plates with 200 µL/well block buffer B for 2 h at 37°C.
4. Wash plates as before.
5. Dilute serum samples 1:50 in wash buffer containing 2% BSA and add 100 µL/well to plates. Incubate for 1 h at 37°C.
6. Wash plates as before.
7. Add 100 µL/well of AP-conjugated antihuman IgG γ-chain-specific antibody diluted 1:1000 in wash buffer containing 2% BSA.
8. Wash plates as before.
9. Add 100 µL/well of the AP substrate and allow the color to develop in the dark at room temperature for 10–60 min. Add stop solution (50 µL/well).
10. Measure the absorbance at 405 nm and standardize the assay against a positive control as described in **Subheading 3.2.1.**

3.2.3. Interpretation of Results of Assays for C1qCLR

In order to determine whether a patient is positive for IgG, anti-C1qCLR antibodies it is necessary to establish the level for a positive result. For the two tests described above, this was achieved by statistically analyzing the

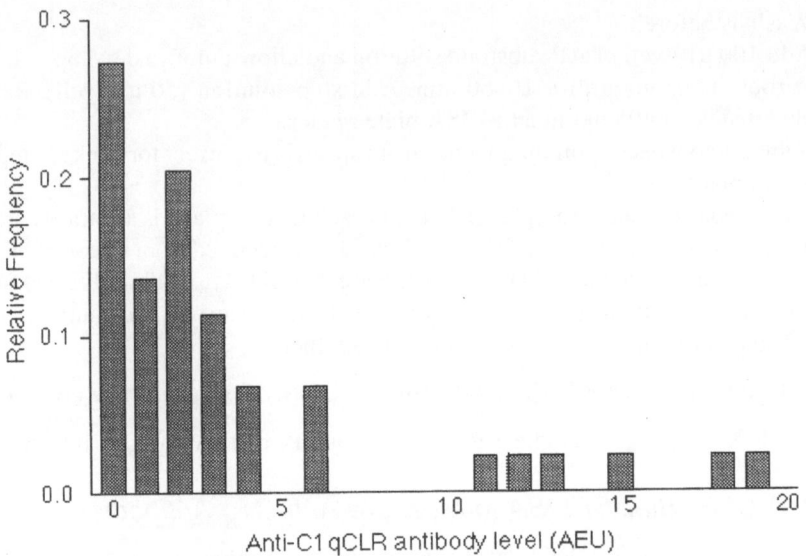

Fig. 2. Histogram illustrating the distribution of IgG anti-C1qCLR antibodies in a population of 44 normal subjects. From this distribution, a positive cut was set at 20 arbitrary ELISA units (mean + 3 standard deviations) and a strongly positive sample was defined as falling above 31 arbitrary ELISA units (mean + 5 standard deviations).

distribution of results in 45 normal serum samples. The cutoff was then set at 20 arbitrary ELISA units for a positive result (mean + 3 standard deviations) and 31 arbitrary ELISA units for a strongly positive result (mean + 5 standard deviations (*see* **Fig. 2**).

3.3. Measurement of Autoantibodies Against C1-Inh (25)

1. Coat 96-well plates overnight at 4°C with 100-μL purified C1-Inh (5 μg/mL) in coating buffer.
2. Wash plates three times in wash buffer.
3. Block plates for 1 h at room temperature with 100 μL/well block buffer B.
4. Wash plates as before.
5. Add 100 μL/well of patient serum diluted serially in wash buffer, incubate for 2 h at room temperature.
6. Wash plates as before.
7. Add 100 μL/well of HRP-conjugated antihuman Ig (recognizing IgA, IgG, or IgM), and incubate for 1 h at room temperature (*see* **Note 2**).
8. Wash plates as before
9. Add 100 μL/well HRP substrate and allow to develop in the dark.
10. Stop the reaction at a suitable point by the addition of stop solution, 50 μL/well.

Table 1
Reagent Mixtures for Varying the Percentage
of Acrylamide in SDS-PAGE Resolving Gels

Reagent	Percentage of acrylamide					
	6%	8%	10%	12.5%	15%	17.5%
Acrylamide	6 mL	8 mL	10 mL	12.5 mL	15 mL	17.5 mL
Bis-acrylamide	0.75 mL	1.05 mL	1.35 mL	1.65 mL	2 mL	2.3 mL
Tris pH 8.7	11.2 mL	11.2 mL	11.2 mL	11.2 mL	11.2 mL	11.2 mL
SDS 20%	150 µL	150 µL	150 µL	150 µL	150 µL	150 µL
ddH$_2$O	11.8 mL	9.6 mL	7.3 mL	4.5 mL	1.7 mL	0 mL
TEMED	25 µL	25 µL	25 µL	25 µL	25 µL	25 µL
10% APS	100 µL	100 µL	100 µL	100 µL	100 µL	100 µL

11. Read the optical density using an automated ELISA plate reader.
12. Results are analyzed essentially as described in **Subheading 3.2.1.**

3.4. SDS-PAGE and Western Blotting

SDS-PAGE originally devised by Laemmli in 1970, is a widely used technique that separates proteins according to their size and mobility in acrylamide gels *(42)*. Like ELISA, Western blotting is a commonly practiced technique to screen for the presence of antibodies to any protein of interest. The protein is usually purified before it is resolved on a gel, however, in certain circumstances whole serum may be used if it is a fluid-phase molecule, or a crude cell lysate if it is a cell bound or intracellular protein. In these cases, the protein can be recognized by the position of the band in comparison to molecular-weight standards. The molecule may also be reduced to its individual subunits that may allow the investigator to further characterize the binding of the antibody in question.

3.4.1. Preparing and Running SDS-PAGE Gels

1. Resolving gels are prepared from stock solutions containing 30% (w/v) acrylamide, 2% (w/v) N,N' methylene *bis*-acrylamide, 20% (w/v) SDS, 1 *M* Tris/HCl pH 8.7, and ddH$_2$O. The ratios of the solutions can be varied (*see* **Table 1**) to alter the final acrylamide concentration from 8% to 15%, according to the size of the molecules to be resolved. *See* **Note 1**.
2. Polymerization of acrylamide is initiated with 0.05% TEMED, 0.02% ammonium persulphate, and the gel mixtures immediately poured into the gel cassette. Polymerization is allowed to proceed for 1 h at room temperature.
3. A stacking gel comprising 0.127 *M* Tris/HCl pH 6.9, 4% (w/v) acrylamide, 0.1% SDS, 0.25% TEMED, and 0.1% ammonium persulphate is then added to the top of the resolving gel.

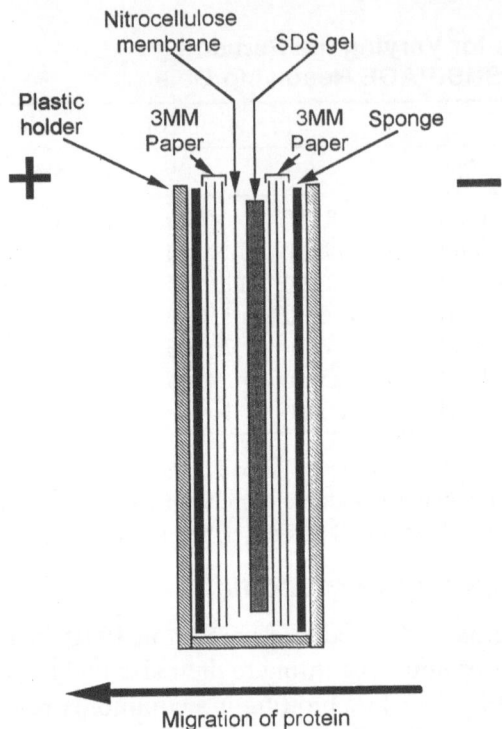

Fig 3. Diagram of Western blot transfer gel "sandwich."

4. A comb is inserted into the stacking gel to generate wells into which samples may be loaded. The stacking gel is allowed to set for 20 min and the comb is then removed and the wells flushed with running buffer.
5. Protein-containing samples are mixed with an equal volume of loading buffer and boiled briefly before loading on the gel. Include molecular weight markers so that the size of the protein band can be estimated.
6. The gel is then placed in the electrophoresis tank filled with running buffer and samples are then resolved by electrophoresis (*see* **Note 3**).

3.4.2. Western Blotting of SDS-PAGE Gels

1. The gel is soaked for 20 min in transfer buffer and then placed in a sandwich against a nitrocellulose filter and 3MM filter sheets that have also been presoaked in the transfer buffer (*see* **Fig. 3**).
2. This sandwich is then placed in an electrophoresis tank filled with transfer buffer. The transfer is initiated by passing an electric current across the sandwich so the protein moves from the gel onto the membrane. This is usually carried out at 4°C at either 250 mA for 3–4 h or 100 mA overnight, but will vary according

to the electrophoretic mobility of the protein in question and the precise equipment used.

3. The membrane is removed from the cell and placed in blocking buffer either overnight at 4°C or for 3 h at room temperature. Blocking is usually achieved in a sealed bag using end-over-end rotation making sure all the air is expelled before closing the bag, as this may leave gaps in the blocking of the membrane (*see* **Note 4**).

4. The blocked membrane is now ready for probing with a relevant autoantibody. This is usually achieved by diluting patient serum or purified antibody in the range of 1:1000 in wash buffer, but a titration may be required according to the concentration of the antibody. As with blocking, probing is accomplished in a sealed bag with end-over-end rotation. Incubate for 2 h at room temperature. Wash three times in wash buffer.

5. HRP-conjugated antihuman Ig diluted 1:1000 (titration may be required to establish the optimal dilution) in wash buffer is added. Incubate for 2 h at room temperature using end-over-end rotation. The isotype and subclass specificity of the secondary antihuman immunoglobulin can be varied according to the Ig being screened.

6. Blots are then washed and developed using an appropriate HRP substrate.

3.5. Identification of Nephritic Factors

Assays for C3 NeF depend on the fact that treatment of human serum with EDTA chelates both Ca^{2+} and Mg^{2+} ions and abolishes both classical (Ca^{2+}-dependent) and alternative (Mg^{2+}-dependent) pathway breakdown of C3. However, addition of EGTA and Mg^{2+} ensures that C3 breakdown can only occur via the AP. Assays detect the presence of C3 breakdown products. Numerous protocols have been published. Here, we describe a two-dimensional electrophoresis method that resembles a commercially available kit (Dakopatts, Copenhagen, Denmark), and a rocket electrophoresis method, both of which detect C3d.

C4 NeF is more difficult to assay and a convertase stabilization assay is widely employed, which makes use of the fact that C4 NeF is capable of prolonging the half-life of both the fluid- and solid-phase classical pathway C3 convertase *(38)*. For more specific details about assays for C4 NeF, we advise consultation with some of the primary references, as our laboratory has no first-hand experience in measuring this particular anticomplement autoantibody *(34–39)*.

3.5.1. Two Dimensional Electrophoresis for Identification of C3 NeF

1. 1.2% agarose dissolved in GBT buffer at 60°C is poured evenly onto 8 × 8-cm glass plates (10 mL/plate) and allowed to set at 4°C in a wet box to avoid the agarose drying out.

2. Three evenly spaced rows each containing three 3-mm wells are then punched out of the gel.

3. 5 µL of normal serum is mixed with 5 µL of patient serum and 1 µL of either PBS, 0.2 *M* EDTA or PBS containing 0.2 *M* EGTA and 0.1 *M* MgCl$_2$. The tubes are incubated at 37°C for 30 min.

4. 7.5 µL from each of the tubes is then added to one of the 3-mm wells and the gel is electrophoresed at 60 V for 3 to 3.5 h at 4°C in a directional parallel to the rows of wells (*see* **Note 5**).

5. The three rows of wells are cut out in 1-cm-wide strips and spaced out on a clean 8 × 8-cm glass plate.

6. 3 mL from a 10 mL stock of molten 1.2% agarose at 56°C containing a 1:100 dilution of a polyclonal anti-C3 antibody is poured in between each strip using a warmed glass pipet (*see* **Note 5**).

7. These plates are then allowed to set at 4°C in a wet box as before.

8. The gels are then electrophoresed in the same cooled (4°C) flatbed tank overnight at 40 V so that the protein in the strips moves into the newly poured agarose containing the anti-C3 antibody.

9. The next morning, the gels are washed four times in PBS over 2 h and then transferred to Gel Bond™ (FMC Bioproducts) and dried.

10. The immobilized, dried gels are stained with 0.5% Coomassie Brilliant Blue R250 in 45% methanol/10% acetic acid for 30 min to 1 h.

11. The peaks are then visualized by destaining the gel with 45% methanol/10% acetic acid and drying. A typical example of this assay is shown in **Fig. 4**, clearly illustrating the layout of the second electrophoresis plate and the two peaks containing the larger and smaller C3 breakdown products that this assay is designed to detect.

3.5.2. Double-Decker Rocket Electrophoresis for Identification of C3 NeF

1. 5.4 mL of melted 1% agarose at 56°C in 74 m*M* Tris, 24 m*M* barbital, pH 8.6, containing a 1:100 dilution of the polyclonal anti-C3c antibody is mixed and poured onto a 10 × 10-cm glass plate so that it fills the bottom 3 cm of the plate (*see* **Fig. 5**). A barrier must be paced across the plate to prevent the agarose spreading further—we use a second glass plate for this purpose.

2. Allow gel to set at 4°C in wet box. Remove barrier.

3. 9 mL of melted agarose gel at 56°C in the above buffer and containing a 1:100 dilution of the anti-C3d antibody is then poured onto the top part of the glass plate (*see* **Fig. 5**).

4. Allow gel to set as before.

5. The samples are prepared as described for **stage 3** in **Subheading 3.5.1.**

6. To generate a C3d standard curve, dilutions of fully activated normal human pooled serum are used (*see* **Note 6**).

7. 2-mm wells are punched in the anti-C3c-containing gel (*see* **Fig. 5**).

8. 2-µL samples are added to the wells and electrophoresis is performed for 18 h at 2.5 V/cm.

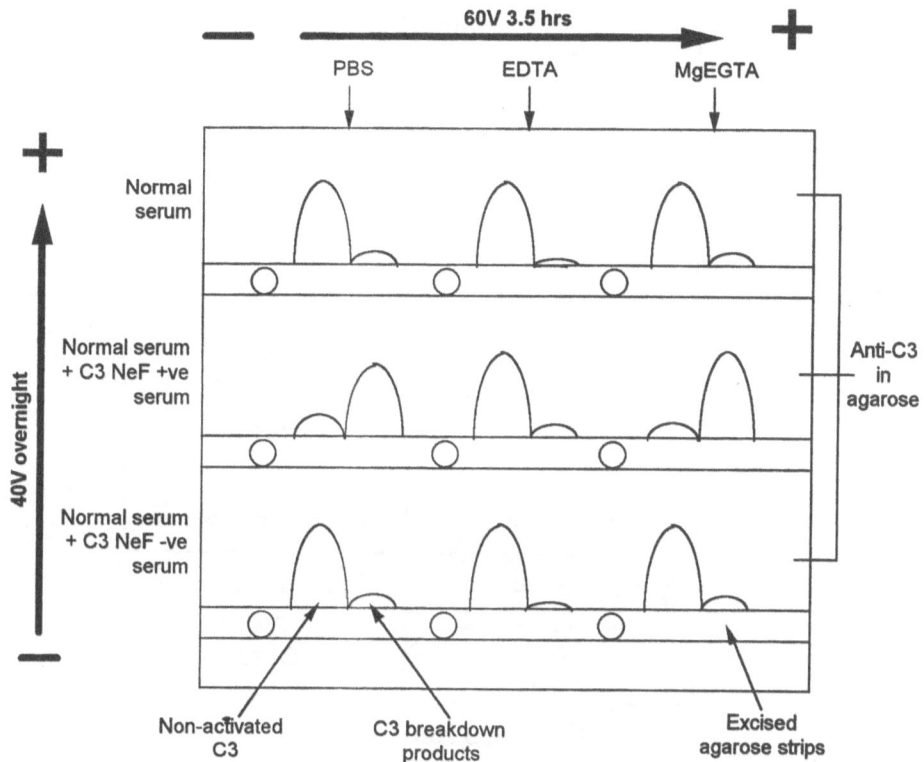

Fig 4. Diagramatic representation of a two-dimensional electrophoresis assay to detect a C3 NeF. The first electrophoresis is shown from left to right and the second from bottom to top. A positive control for this assay consists of serum from a previously identified C3 NeF positive patient. A negative control is normal pooled human serum that has not been mixed with any patient's serum.

9. The gels are then pressed, dried and stained as described in **stages 8–10** in **Subheading 3.5.1.**
10. The height of the rockets is measured from a set point, usually the middle of the sample well. The concentration of C3d present is then estimated by extrapolation from the C3d standard curve.

4. Notes

1. Good levels of laboratory practice must be maintained when handling patient sera, as these are potential sources of infectious material. Care must also be taken in the use of any hazardous chemical substances required for the individual techniques. Acrylamide is a cumulative neurotoxin that is absorbed through skin so the user should strongly adhere to the recommended safety precautions.

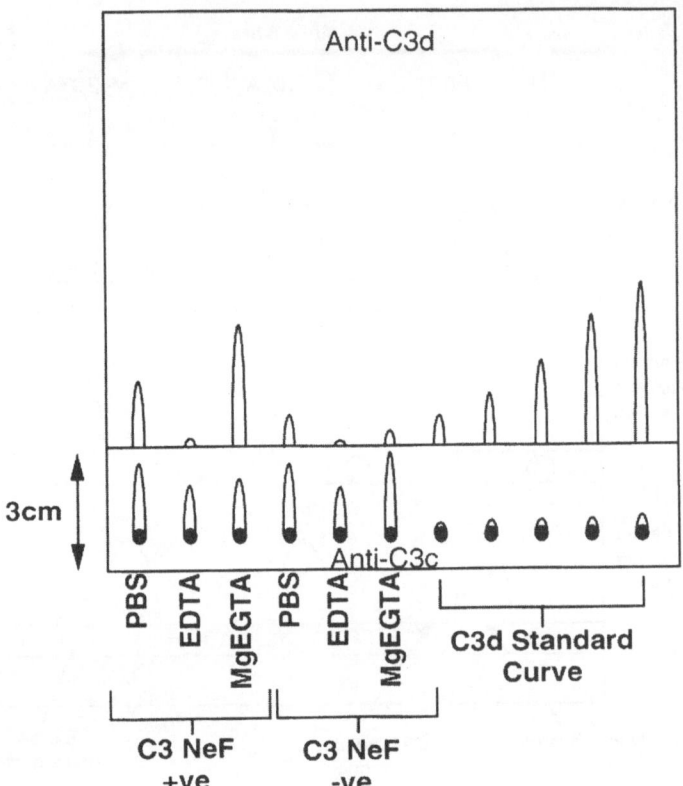

Fig 5. Double-decker rockets for the detection of C3 NeF.

2. Sera containing anti-C1-Inh antibodies can be analyzed using peroxidase-conjugated antibodies to specific isotypes (IgA, IgG, and IgM) and individual light chains (κ and λ).

3. When electrophoresing proteins, remember that they are negatively charged and therefore migrate from the negative cathode to the positive anode. The current, voltage, and time taken to resolve the fragments will depend upon the equipment used.

4. Always handle the nitrocellulose membrane with gloves and then only at the edges to avoid contamination. There are different nitrocellulose membranes for nucleic acids and protein eg N+ for nucleic acids and C for proteins available from Amersham Pharmacia.

5. The first electrophoresis step must be long enough to separate the two peaks of C3 breakdown products. The amount of anti-C3 antibody in the second electrophoresis step will need titrating to give good size peaks. Too much antibody and the peaks (precipitin lines) are too small, too little antibody and the peaks become faint and difficult to visualize. It is essential that the anti-C3 antibody is

polyclonal and can recognize epitopes on both C3b and the smaller C3 break-down products (e.g., C3d and C3dg) to give two peaks.

6. To generate fully activated serum, pooled normal serum containing 15 mM NaN$_3$ is incubated at 37°C for 5 d to cause complete activation of C3. The amount of C3d generated is taken as approx 1000 µg/L. This activated serum can then be aliquoted and stored at –70°C until required. When required the pooled activated serum is diluted in PBS containing 15 mM EDTA to give a standard curve in the range of 10–200 µg/L.

References

1. Antes, U., Heinz, H. P., and Loos, M. (1988) Evidence for the presence of autoantibodies to the collagen-like portion of C1q in systemic lupus erythematosus. *Arth. Rheum.* **31,** 457–464.
2. Golan, M. D., Burger, R., and Loos, M. (1982) Conformational changes in C1q after binding to immune complexes: detection of neoantigens with monoclonal antibodies. *J. Immunol.* **129,** 445–447.
3. Hoekzema, R., Martens, M., Brouwer, M. C., and Hack, C. E. (1988) The distortive mechanism for the activation of complement component C1 supported by studies with a monoclonal antibody against the "arms" of C1q. *Mol. Immunol.* **25,** 485–494.
4. Wener, M. H., Uwatoko, S., and Mannik, M. (1989) Antibodies to the collagen-like region of C1q in sera of patients with autoimmune rheumatic diseases. *Arth. Rheum.* **32,** 544–551.
5. Prada, A. E. and Strife, C. F. (1992) IgG subclass restriction of autoantibody to solid-phase C1q in membranoproliferative and lupus glomerulonephritis. *Clin. Immunol. Immunopathol.* **63,** 84–88.
6. Strife, C. F., Leahy, A. E., and West, C. D. (1989) Antibody to a cryptic, solid phase C1Q antigen in membranoproliferative nephritis. *Kidney Int.* **35,** 836–842.
7. Siegert, C. E., Daha, M. R., van der Voort, E. A., and Breedveld, F. C. (1990) IgG and IgA antibodies to the collagen-like region of C1q in rheumatoid vasculitis. *Arth. Rheum.* **33,** 1646–1654.
8. Marder, R. J., Potempa, L. A., Jones, J. V., Toriumi, D., Schmid, F. R., and Gewurz, H. (1984) Assay, purification and further characterization of 7S C1q-precipitins (C1q-p) in hypocomplementemic vasculitis urticaria syndrome and systemic lupus erythematosus. *Acta Pathol. Microbiol. Immunol. Scand. Suppl.* **284,** 25–34.
9. Wisnieski, J. J. and Naff, G. B. (1989) Serum IgG antibodies to C1q in hypocomplementemic urticarial vasculitis syndrome. *Arth. Rheum.* **32,** 1119–1127.
10. Siegert, C. E., Daha, M. R., Halma, C., van der Voort, E. A., and Breedveld, F. C. (1992) IgG and IgA autoantibodies to C1q in systemic and renal diseases. *Clin. Exp. Rheumatol.* **10,** 19–23.
11. Coremans, I. E., Daha, M. R., van der Voort, E. A., Muizert, Y., Halma, C., and Breedveld, F. C. (1992) Antibodies against C1q in anti-glomerular basement membrane nephritis. *Clin. Exp. Immunol.* **87,** 256–260.

12. Siegert, C., Daha, M., Westedt, M. L., van der Voort, E., and Breedveld, F. (1991) IgG autoantibodies against C1q are correlated with nephritis, hypocomplementemia, and dsDNA antibodies in systemic lupus erythematosus. *J. Rheumatol.* 18, 230–234.

13. Uwatoko, S., Aotsuka, S., Okawa, M., Egusa, Y., Yokohari, R., Aizawa, C., and Suzuki, K. (1987) C1q solid-phase radioimmunoassay: evidence for detection of antibody directed against the collagen-like region of C1q in sera from patients with systemic lupus erythematosus. *Clin. Exp. Immunol.* **69,** 98–106.

14. Siegert, C. E., Daha, M. R., Tseng, C. M., Coremans, I. E., van Es, L. A., and Breedveld, F. C. (1993) Predictive *Ann. Rheum. Dis.* value of IgG autoantibodies against C1q for nephritis in systemic lupus erythematosus. *Ann. Rheum. Dis.* **52,** 851–856.

15. Coremans, I. E., Spronk, P. E., Bootsma, H., Daha, M. R., van der Voort, E. A., Kater, L., Breedveld, F. C., and Kallenberg, C. G. (1995) Changes in antibodies to C1q predict renal relapses in systemic lupus erythematosus. *Am. J. Kidney Dis.* **26,** 595–601.

16. Norsworthy, P., Theodoridis, E., Botto, M., Athanassiou, P., Beynon, H., Gordon, C., et al. (1999) Over-representation of the FcγRIIA R131/131 genotype in caucasoid SLE patients with auto-antibodies to C1q and glomerulonephritis. *Arth. Rheum.* **42,** 1828–1832.

17. Wisnieski, J. J., Baer, A. N., Christensen, J., Cupps, T. R., Flagg, D. N., Jones, J. V., et al. (1995) Hypocomplementemic urticarial vasculitis syndrome. Clinical and serologic findings in 18 patients. *Med. Balt.* **74,** 24–41.

18. Davies, A. E. (1988) C1 inhibitor and hereditary angioneurotic edema. *Ann. Rev. Immunol.* **6,** 595–628.

19. Cicardi, M., Beretta, A., Colombo, M., Gioffre, D., Cugno, M., and Agostoni, A. (1996) Relevance of lympho proliferative disorders and of anti-C1 inhibitor autoantibodies in acquired angio-oedema. *Clin. Exp. Immunol.* **106,** 475–480.

20. Geha, R. S., Quinti, I., Austen, K. F., Cicardi, M., Sheffer, A., and Rosen, F. S. (1985) Acquired C1–inhibitor deficiency associated with antiidiotypic antibody to monoclonal immunoglobulins. *N. Engl. J.* Med. **312,** 534–540.

21. Malbran, A., Hammer, C. H., Frank, M. M., and Fries, L. F. (1988) Acquired angioedema: observations on the mechanism of action of autoantibodies directed against C1 esterase inhibitor. *J. Aller. Clin. Immunol.* **81,** 1199–1204.

22. Mandle, R., Baron, C., Roux, E., Sundel, R., Gelfand, J., Aulak, K., et al. (1994) Acquired C1 inhibitor deficiency as a result of an autoantibody to the reactive center region of C1 inhibitor. *J. Immunol.* **152,** 4680–4685.

23. He, S., Tsang, S., North, J., Chohan, N., Sim, R. B., and Whaley, K. (1996) Epitope mapping of C1 inhibitor autoantibodies from patients with acquired C1 inhibitor deficiency. *J. Immunol.* **156,** 2009–2013.

24. Alsenz, J., Bork, K., and Loos, M. (1987) Autoantibody-mediated acquired deficiency of C1 inhibitor. *N. Engl. J. Med.* **316,** 1360–1366.

25. Spitzer, R. E., Vallota, E. H., Forristal, J., Sudora, E., Stitzel, A., Davis, N. C., and West, C. D. (1969) Serum C'3 lytic system in patients with glomerulonephritis. *Science* **164,** 436–437.

26. Peters, D. K., Charlesworth, J. A., Sissons, J. G., Williams, D. G., Boulton Jones, J. M., Evans, D. J., et al. (1973) Mesangiocapillary nephritis, partial lipodystrophy, and hypocomplementaemia. *Lancet* **2,** 535–538.

27. Vallota, E. H., Forristal, J., Spitzer, R. E., Davis, N. C., and West, C. D. (1970) Characteristics of a non-complement-dependent C3-reactive complex formed form factors in nephritic and normal serum. *J. Exp. Med.* **131,** 1306–1324.

28. Vallota, E. H., Forristal, J., Spitzer, R. E., Davis, N. C., and West, C. D. (1971) Continuing C3 breakdown after bilateral nephrectomy in patients with membranoproliferative glomerulonephritis. *J. Clin. Invest.* **50,** 552–558.

29. Daha, M. R., Fearon, D. T., and Austen, K. F. (1976) C3 nephritic factor (C3NeF): stabilization of fluid phase and cell-bound alternative pathway convertase. *J. Immunol.* **116,** 1–7.

30. Weiler, J. M., Daha, M. R., Austen, K. F., and Fearon, D. T. (1976). Control of the amplification convertase of complement by the plasma protein beta1H. *Proc. Natl. Acad. Sci. USA* **73,** 3268–3272.

31. Tanuma, Y., Ohi, H., and Hatano, M. (1990). Two types of C3 nephritic factor: properdin-dependent C3NeF and properdin-independent C3NeF. *Clin. Immunol. Immunopathol.* **56,** 226–238.

32. Ohi, H., Watanabe, S., Fujita, T., and Yasugi, T. (1992) Significance of C3 nephritic factor (C3NeF) in non-hypocomplementaemic serum with membranoproliferative glomerulonephritis (MPGN). *Clin. Exp. Immunol.* **89,** 479–484.

33. Daha, M. R. and van Es, L. A. (1981) Stabilization of homologous and heterologous cell-bound amplification convertases, C3bBb, by C3 nephritic factor. *Immunology* **43,** 33–38.

34. Halbwachs, L., Leveille, M., Lesavre, P., Wattel, S., and Leibowitch, J. (1980) Nephritic factor of the classical pathway of complement: immunoglobulin G autoantibody directed against the classical pathway C3 convertase enzyme. *J. Clin. Invest.* **65,** 1249–1256.

35. Daha, M. R., Hazevoet, H. M., Vanes, L. A., and Cats, A. (1980) Stabilization of the classical pathway C3 convertase C42, by a factor F-42, isolated from serum of patients with systemic lupus erythematosus. *Immunology* **40,** 417–424.

36. Fujita, T., Sumita, T., Yoshida, S., Ito, S., and Tamura, N. (1987) C4 nephritic factor in a patient with chronic glomerulonephritis. *J. Clin. Lab. Immunol.* **22,** 65–70.

37. Tanuma, Y., Ohi, H., Watanabe, S., Seki, M., and Hatano, M. (1989) C3 nephritic factor and C4 nephritic factor in the serum of two patients with hypocomplementaemic membranoproliferative glomerulonephritis. *Clin. Exp. Immunol.* **76,** 82–85.

38. Gigli, I., Sorvillo, J., and Halbwachs Mecarelli, L. (1985) Regulation and deregulation of the fluid-phase classical pathway C3 convertase. *J. Immunol.* **135,** 440–444.

39. Ohi, H. and Yasugi, T. (1994) Occurrence of C3 nephritic factor and C4 nephritic factor in membranoproliferative glomerulonephritis (MPGN). *Clin. Exp. Immunol.* **95,** 316–321.

40. Sakari Jokiranta, T., Solomon, A., Zipfei, P. F., Pangburn, M. K., and Meri, S. (1996) Structure and function of a nephritogenic λ-light chain dimer—a unique human miniautoantibody against factor H. *Mol. Immunol.* **33,** 11.
41. He, H. and Lin, Y-L. (1998) In vitro stimulation of C1s-presenting autoantibodies from patients with systemic lupus erythematosus. *J. Immunol.* **160,** 4641–4647.
42. Laemmli, U. K. (1970) Cleavage of structural proteins during the assembly of the head of bacteriophage T4. *Nature* **227,** 680–686.

14

Allotyping of Complement Components

Reinhard Würzner

1. Introduction

Studies on the genetics of complement proteins were originally initiated by the discovery of complement deficiencies in animals and man. From these studies, it has become apparent that most complement components are polymorphic, some with several tens of different allotypes. The study of the major histocompatibility complex (MHC)-encoded complement components C4A and C4B was particularly rewarding because of their association with human lymphocyte antigens (HLA) and because of their presence as closely linked loci, both of which encode for a large number of allelic variants (1).

Complement genetics has also been a valuable tool to investigate plasma protein genetics in general and their evolution. The chromosomal assignment of the genes coding for complement proteins reveals interesting linkage groups of structurally homologous components (**Fig. 1**). These groupings confirm previous assumptions, based on homology studies at the protein level, that the majority of complement proteins have evolved by duplication from only a small number of precursor genes (2,3).

Because complement receptors and certain regulatory proteins are also polymorphic and are expressed on erythrocytes, they have the potential to represent blood group antigens: the Knops, McCoy, Swain-Langley, and York antigens are known to be on CR1. Variations in the DAF antigen are responsible for the Cromer blood group system with the rare Inab phenotype lacking DAF altogether. Chido and Rodgers blood group antigens are associated with C4. In this respect, complement genetics has been widely applied to both forensic medicine and population genetics (4).

With the advent of recombinant DNA methodology, immense progress in new technologies has facilitated the characterization of complement allotypes

From: *Methods in Molecular Biology, vol. 150: Complement Methods and Protocols*
Edited by: B. P. Morgan © Humana Press Inc., Totowa, NJ

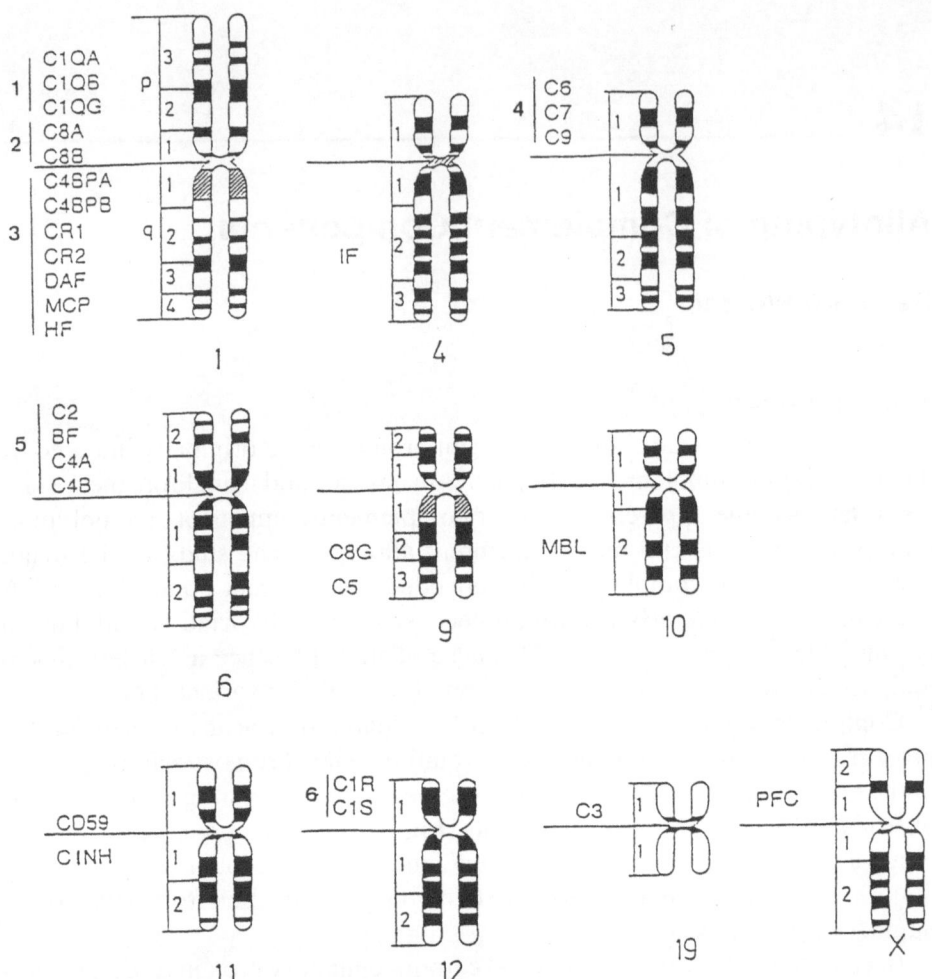

Fig. 1. Schematic diagram of the location of structural genes of complement proteins, indicated on the left. Six clusters can be observed: (1) C1Q gene cluster; (2) C8A/B gene cluster; (3) regulators of complement activation (RCA) gene cluster (all on chromosome 1); (4) terminal complement complex (TCC) or membrane attack complex (MAC) gene cluster (on chromosome 5); (5) major histocompatibility complex (MHC) class III gene cluster (on chromosome 6); and (6) C1R/S gene cluster (on chromosome 12). Genes for complement related proteins (e.g., surfactant proteins, perforin, vitronectin) or complement receptors are not depicted.

on the molecular level. Most genes coding for complement components, control proteins, receptors or related proteins have been cloned, and at least partially sequenced. Thus, polymorphism studies that were restricted to phenotypic analyses of the respective protein can now be also assessed at the genomic level. Both phenotypic assessments of variability at the protein level (phenotyping) and characterization of genomic DNA (genotyping) are currently used *(5)*, as both have their advantages and disadvantages as discussed below.

1.1. Phenotyping

Phenotyping is traditionally performed using methods that analyze the electrophoretic mobility or isoelectric point of proteins in agarose or polyacrylamide gel electrophoresis. These separation techniques rely on charge and size/ conformation differences or on charge differences alone. Detection is achieved by (1) direct nonspecific protein staining, (2) by immunofixation, (3) by electro- or press blotting procedures followed by immunological detection, or (4) functional (hemolytic) overlay methods using hemolytic intermediates. Details of sodium dodecyl sulfate-polyacrylamide gel electrophoresis (SDS-PAGE) separation and western blotting are given in Chapter 13 and the use of hemolytic overlays is described in Chapter 11. The latter technique has the disadvantage that it is less sensitive and dependent on the time of development, but has the important advantage that all bands detected represent functionally/ hemolytically active protein. Very detailed protocols for the phenotyping by isoelectric focusing (IEF) of C4, BF, C3, C6, and C8 have recently been published elsewhere in "cook book format" *(2,5)*. Interesting modifications, such as the use of neuraminidase for the pretreatment of samples and taurine as an additive to the gels are explained there in detail.

Some complement receptors exhibit a structural polymorphism (molecular weight, e.g., CR1) immediately apparent on gel electrophoresis or a functional polymorphism affecting the number of molecules expressed per cell which is commonly determined by flow cytometry *(6)*. The allotypes of the complement related blood group antigens, mentioned above, are assessed serologically *(4)*.

The use of monoclonal antibodies for complement allotyping is a relatively novel and very sensitive technique that offers considerable advantages, but has yet to gain wide acceptance. These methods will be described in detail in this chapter.

1.2. Genotyping

Genotyping may be achieved by studying component-specific DNA restriction fragment length polymorphisms (RFLPs), even when the location of the variation remains unknown. In situations where the site of the polymorphism has already been mapped, genotyping can be performed by polymerase chain reaction (PCR) using specific primers followed by enzymatic digestion or sequencing or by using allele-specific primers, usually in a nested PCR. A number of the phenotypically common variants, but by no means all expressed allotypes, have been identified in DNA sequences as nucleotide exchanges, deletions, or insertions. Thus, they can be assessed using either phenotyping or genotyping.

A number of variations have, however, been found in noncoding parts (introns) of a gene, usually without any functional consequences for the protein. Other silent polymorphisms have been found in exons: in this case, electrically neutral amino acid exchanges or nucleotide, but not amino acid exchanges are responsible. Variants have been revealed that are phenotypically indistinguishable from, and thus grouped with, known allotypes when assessed by conventional phenotyping techniques, but nevertheless, show a different variation upon genotyping. Thus, some phenotypic (IEF) allotypes can be further subdivided using genotyping. Very detailed protocols for genotyping complement components have recently been published elsewhere *(2,5)*.

Genotyping has the disadvantage that sequence variations are not confined to short stretches of genomic DNA, but are rather scattered throughout the entire gene. Therefore, any conclusions drawn from PCR results only correspond to a limited stretch of DNA, whereas (one single) isoelectric focusing assesses the entire protein.

Another disadvantage is that, in contrast to all phenotyping methods, neither the presence nor the functional activity of the protein encoded at a particular allele can be ascertained. In this respect, genotyping alone does not allow identification of nonexpressed ("silent" or "null") alleles. However, once the mutation is known, the defective gene may be traced in family studies, providing a basis for genetic counselling for the affected families.

One of the great advantages of genotyping is its efficiency: well-defined PCR detection allows the screening of a large number of samples, either simultaneously or as pooled samples. Thus, genotyping has not replaced phenotyping (also see below) and a combination of both is usually applied to obtain reliable typing data from complex polymorphic proteins *(7,8)*.

1.3. Allotyping Using Allospecific Monoclonal Antibodies (MAbs)

Allospecific MAbs are classified by their ability to exclusively react specifically with one or more defined allotypes of a particular component. They have been characterized for C4, C3, and C7 and are usually used in Western blot or enzyme-linked immunosorbent assay (ELISA) procedures.

The Rodgers 1,2 specific MAbs 99H7 and VlaD2 recognize most C4A allotypes, but also some C4B allotypes, whereas the Chido 1 specific MAbs 2B12, 1217, and 1228 are reactive with most C4B allotypes, but also some C4A allotypes—thus, both groups of MAbs are allotype-specific for a panel of allotypes *(9)*. More recently, antipeptide MAbs were obtained from mice immunized with

C4A- or C4B-specific peptides containing the discordant amino acid residues 1101–1106. These MAbs AII 1 and BII 1 react exclusively with C4A or C4B allotypes, respectively *(10)*.

Additional protein polymorphisms based on the reactivity of allotype specific monoclonal antibodies have been characterized for C3 and C7. In both cases, the allotype is determined by a proline residue that causes the loss of the epitope detected by the MAb, whereas leucine (in C3) or threonine (in C7) at the same position gives an allotype recognized by the respective MAbs *(11,12)*. The allospecific anti-C3 antibody (HAV 4-1) was discovered via its reaction pattern on immunoblots *(13)*, whereas the allospecific anti-C7 antibody (WU 4-15) was found by comparing its reaction pattern in ELISA with that of polyclonal anti-C7 IgG, in a protocol where both were used as coating antibody *(14,15)*. C7 M/N typing has been successfully used to trace the source of C7 after liver transplantation *(16,17)* showing that the liver is not the exclusive site of synthesis. It has also been widely used in terminal complement deficiency studies to identify homozygous subjects or unaffected siblings or trace heterozygous carriers of C7 defects, and, because of the genetic linkage with the MAC gene cluster, also of C6 and C9 defects *(18,19)*. In particular, C7 M/N typing has proven useful in determining the cause of phenotypic complete C7 deficiency in a Russian subtotal C7D subject with anomalous response to transfusion therapy. The source of C7 appearing after transfusion was ascribed to the patient himself, and not to the transfused material *(20)*, demonstrating his ability to synthesize low amounts of the C7 protein. C7 M/N typing was also used to show that the bone marrow markedly contributes to the circulating C7 level *(17)* and that the C7 released by PMNs is, at least in vitro, not taken up from the serum, but presumably synthesized by a bone marrow progenitor *(21)*. The assay has also been applied to anthropological studies, as reviewed *(22)*, which demonstrated that, as for many other complement proteins, the allele frequencies differ considerably between different populations. Recently, C7 M/N typing added further evidence to the existence of duplicated C7 genes in some (very few) subjects *(19)*. The assay protocol is described in detail below.

Numerous associations between "silent" or "null" alleles of complement components and disease have been reported as detailed in Chapter 11. Disease associations with expressed complement alleles are less clear. An increased prevalence of the C3F allele has been reported in patients with partial lipodystrophy, IgA nephropathy, or Indian childhood hepatic cirrhosis *(23)*, and the C7 alleles C7*3 and C7*N have been shown to represent hypomorphic alleles *(19)*. Nevertheless, it really appears doubtful whether any complement allele, not causing functional or immunochemical deficiency, is associated with disease.

For the alleles of complement components coded within the MHC class III gene cluster (**Fig. 1**), it still has to be demonstrated that the reported associations with disease are really due to the altered complement protein, and not because of the linkage to a particular HLA antigen.

2. Materials

1. Spectrophotometer capable of reading optical densities in 96-well microtiter plates at 405 or 410 nm and 490 nm (reference filter).
2. Optional: automatic washer for 96-well microtiter plates.
3. Optional: multichannel pipet.
4. 96-flat bottom well microtiter plates (Titertek™ plates from Flow Laboratories).
5. Appropriate laboratory gloves.
6. Phosphate buffered saline (PBS; supplied in tablet form by Oxoid).
7. Coating buffer consisting of 0.2 M sodium carbonate, pH 10.6.
8. Blocking buffer consisting of coating buffer or PBS, supplemented with 0.5% gelatine or bovine serum albumin.
9. Incubation and washing buffer (IWB) consisting of PBS supplemented with 0.05% Tween 20 (Sigma).
10. Standard serum or plasma of allotype C7 M, control sera with known C7 M/N phenotypes.
11. MAb WU 4-15 (for ELISA M) *(14)* and goat anti-C7 (for ELISA P) (coating and detection antibodies and reference samples are available from the author upon request; almost any good anti-C7 IgG may be used, but no similar MAb exists).
12. Goat anti-C7 IgG, biotinylated.
13. Streptavidin-horseradish peroxidase (Boehringer Mannheim, 1089152) or streptavidin-alkaline phosphatase (DAKO, D 0396).
14. Horseradish peroxidase development system: use 0.5-2 mM ABTS (2,2 azino-di(3-ethyl)-benzthiazoline sulphonate, e.g. (Boehringer Mannheim, 102946)) in 0.1 M acetate buffer (alternatively: supplemented with 0.05 M sodium phosphate), pH 4.0–4.2, to which H_2O_2 (2.5 mM, 30 µL of 3% H_2O_2 per 10 mL acetate buffer) is added immediately before use; or alkaline phosphatase development system: dissolve 1 tablet pNPP (p-Nitrophenyl Phosphate, Sigma) in 5 mL substrate buffer, consisting of 0.1 M glycine, 1 mM MgCl$_2$, 1 mM ZnCl$_2$, pH 10.4.
15. Optional: 10% H_2SO_4 (for ABTS) or 3 N NaOH (for pNPP) as stop solution.

3. Methods

1. Coat half of the wells of a 96-well microplate by overnight incubation at 4–8°C (superior to shorter incubation times or higher incubation temperatures) with 100 µL of WU 4–15, 5 µg/mL in coating buffer or PBS (ELISA M) and the other half with goat anti-C7, 20 µg/mL in coating buffer or PBS (ELISA P). Seal the plate with parafilm (American National Can Co.) to reduce evaporation effects (*see* **Note 1**).
2. Remove the coating buffer by deflecting the plate, but do not wash with IWB as the detergent (Tween) will affect the following blocking step. Use 150 µL blocking buffer per well in order to avoid nonspecific adsorption to those parts of the wells that become exposed when the plate is not held or placed exactly horizontally

(e.g., when you carry the plate). Block for about 30 min at room temperature.

3. Wash plate with IWB at room temperature, at least three times between all incubation steps, volume: 200–300 µL per well. Wash by hand using a multichannel pipet or a plastic bottle, or use an automatic washer (*see* **Note 2**).

4. Use gloves from this step onwards, as human samples are handled. Apply sera or plasma 1/2500 diluted in IWB, and standard (C7 M) serum 1/100, 1/200—1/128,000 diluted in IWB and incubate for about 1 h (longer time intervals, even overnight incubation, are also possible) at room temperature. If you have to use virtually undiluted samples (e.g., when testing deficient subjects), add more Tween in a smaller volume of IWB.

5. Wash the plate three times as in **step 3**, but be particularly careful at the first wash, as some wells may contain very high concentrations of the antigen and others very low. Under these circumstances, even a very short exposure of the latter well to higher amounts of antigen from a neighboring may well yield false positive results. In addition, the amount of potentially hazardous agents in the wells is largest at this step.

6. Apply detection antibody goat anti-C7, biotinylated, diluted 20 µg/mL in IWB for about 1 h at room temperature.

7. Wash plate as in **step 3**.

8. Apply conjugate streptavidin-horseradish peroxidase or streptavidin-alkaline phosphatase, diluted according to the manufacturer (usually 1/1000 or higher) in IWB for about 1 h at room temperature.

9. Wash plate as in **step 3**.

10. Add substrate (ABTS or pNPP, 100 µL per well) rapidly and incubate at room temperature for several minutes until the upper standard reaches an optical density of approx 1.5. Add 100 µL of stop solution if immediate measurement is not possible.

11. Read in a spectrophotometer for 96-well microtiter plates at 405/410 nm using 490 nm as reference filter. Calculate ratios for each sample by dividing the C7 concentration obtained with ELISA *M* by the one obtained with ELISA P. *See* **Note 3** for interpretation of results.

4. Notes

1. Do not coat the outer wells of the plate as they usually yield less reliable results even if the plate is always placed in a moisture box as recommended for all steps. Use the plates within 1 wk.

2. An experienced worker will be faster and be just as reliable as an automatic washer, which is advantageous only when many plates are being used.

3. Samples with a ratio between 0.75 and 1.25 define C7 *M* subjects, whereas ratios between 0.35 and 0.65, or less than 0.02 identify C7 MN or C7 N subjects, respectively.

References

1. Mauff, G., Luther, B., Schneider, P. M., Rittner, C., Stradmann-Bellinghausen, B., Dawkins, R., and Moulds, J. M. (1998) Reference typing report for complement component C4. *Exp. Clin. Immunogenet.* **15,** 249–260.

2. Schneider, P. M. and Rittner, C. (1997) Complement genetics, in *Complement—A Practical Approach* (Dodds, A. and Sim, R. B., eds.), Oxford University, Oxford, pp. 165–198.
3. Hobart, M. J. (1998) The evolution of the terminal complement genes: ancient and modern. *Exp. Clin. Immunogenet.* **15**, 235–243.
4. Rittner, C. and Schneider, P. M. (1988) Complement genetics, in *The Complement System* (Rother, K. and Till, G., eds.), Springer, Berlin, Germany, pp. 80–135.
5. Mauff, G. and Würzner, R. (1996) Complement genetics, in *Handbook of Experimental Immunology* (Herzenberg, L. A., Weir, D. M., Herzenberg, L. A., and Blackwell, C., eds.), Blackwell, Cambridge, MA, pp. 75.1–75.50.
6. Moulds, J. M., Brai, M., Cohen, J., Cortelazzo, A., Lin, M., Sadallah, S., et al. (1998) Reference Typing Report for Complement Receptor One (CR1). *Exp. Clin. Immunogenet.* **15,** 291–294.
7. Schneider, P. M., Rittner, C., Mauff, G., and Würzner, R., eds. (1998) VIIth Complement Genetics Workshop and Conference, Proceedings, Mainz, May 21–23, 1998. *Exp. Clin. Immunogenet.* **15,** 197–300.
8. Schneider, P. M. and Würzner, R. (1999) Complement genetics: biological implications of polymorphisms and deficiencies. *Immunol. Today* **20,** 2–5.
9. Doxiadis, G. and Grosse-Wilde, H. (1990) Allotyping by prolonged gel electrophoresis and immunoblotting using monoclonal and polyclonal antibodies. *Complement Inflamm.* **7,** 269–276.
10. Reilly, B. D., Levine, P., Rothbard, J., and Skanes, V. M. (1991) Monoclonal antipeptide antibodies against amino acid residues 1101-1106 of human C4 distinguish C4A from C4B. *Complement Inflamm.* **8,** 33–42.
11. Botto, M., Fong, K. Y., So, A. K., Koch, C., and Walport, M. J. (1990) Molecular basis of polymorphisms of human complement component C3. *J. Exp. Med.* **172,** 1011–1017.
12. Würzner, R., Fernie, B. A., Jones, A. M., Lachmann, P. J., and Hobart, M. J. (1995) Molecular basis of the complement C7 M/N polymorphism: a neutral amino acid substitution outside the epitope of the allospecific monoclonal antibody WU 4-15. *J. Immunol.* **154,** 4813–4819.
13. Koch, C. and Behrendt, N. (1986) A novel polymorphism of human complement component C3 detected by means of a monoclonal antibody. *Immunogen.* **23,** 322–325.
14. Würzner, R., Nitze, R., and Götze, O. (1990) C7*9, a new frequent allele detected by an allotype-specific monoclonal antibody. *Complement Inflamm.* **7,** 290–297.
15. Würzner, R., Hobart, M. J., Orren, A., Tokunaga, K., Nitze, R., Götze O., and Lachmann, P. J. (1992) A novel protein polymorphism of human complement C7 detected by a monoclonal antibody. *Immunogen.* **35,** 398–402.
16. Würzner, R., Joysey. V. C., and Lachmann, P. J. (1994) Complement component C7: assessment of in vivo synthesis after liver transplantation reveals that hepatocytes do not synthesize the majority of human C7. *J. Immunol.* **152,** 4624–4629.
17. Naughton, M. A., Walport, M. J., Würzner, R., Carter, M. J., Alexander, G. J. M., Goldman, J. M., and Botto, M. (1996) Organ-specific contribution to circulating C7 levels by the bone marrow and liver in humans. *Eur. J. Immunol.* **26,** 2108–2112.

18. Würzner, R., Rance, N., Potter, P. C., Hendricks, M. L., Lachmann, P. J., and Orren, A. (1992) C7 M/N protein polymorphism typing applied to inherited deficiencies of human complement proteins C6 and C7. *Clin. Exp. Immunol.* **89**, 485–489.

19. Würzner, R., Witzel-Schlömp, K., Tokunaga, K., Fernie, B. A., Hobart, M. J., and Orren, A. (1998) Reference typing report for complement components C6, C7 and C9, including mutations leading to deficiencies. *Exp. Clin. Immunogenet.* **15**, 268–285.

20. Würzner, R., Platonov, A. E., Beloborodov, V. B., Pereverzev, A. I., Vershinina, I. V., Fernie, B. A., Hobart, M. J., Lachmann, P. J., and Orren, A. (1996) How partial C7 deficiency with chronic and recurrent bacterial infections can mimic total C7 deficiency: temporary restoration of host C7 level following plasma transfusion. *Immunol.* **88**, 407–411.

21. Høgåsen, A. K. M., Würzner, R., Abrahamsen, T. G., and Dierich, M. P. (1995) Human polymorphonuclear leukocytes store large amounts of terminal complement components C7 and C6 which may be released on stimulation. *J. Immunol.* **154**, 4734–4740.

22. Sölder, B. M., de Stefano, G. F., Dierich, M. P., and Würzner, R. (1996) The Cayapa indians of Ecuador: a genetically isolated group with unexpected complement C7 M/N allele frequencies. *Eur. J. Immunogenet.* **23**, 199–203.

23. Figueroa, J. E. and Densen P. (1991) Infectious diseases associated with complement deficiencies. *Clin. Microbiol. Rev.* **4**, 359–395.

15

Complement and Immune Complexes

Julian T. Nash and Kevin A. Davies

1. Introduction

Serum sickness resulting from repeated administration of horse antitoxins was hypothesized to be an immune complex disease by von Pirquet and Schick at the turn of this century *(1)*. Arthus developed the first animal model of immune complex disease by inducing cutaneous vasculitis and inflammation following injection of antigen in immunized rabbits *(2)*. In the 1940s Heidelberger observed that complement inhibited the formation of precipitating antigen–antibody lattices. In the 1950s and 1960s, experimental serum sickness models of glomerulonephritis and vasculitis demonstrated the potential for circulating immune complexes to cause disease *(3,4)*.

The processing of immune complexes is an important function of the immune system. Immune complexes are implicated in the pathogenesis of many autoimmune diseases, such as systemic lupus erythematosus *(5,6)*. Defective processing of immune complexes by the fixed mononuclear phagocytic system may result in the deposition of immune complexes in many tissues, causing inflammation, the release of autoantigens, and the stimulation of an autoimmune response. Processing of immune complexes may also be important in the localization of antigen to specific sites in the immune system where antigen presentation may occur leading to the generation of an adaptive humoral immune response *(7,8,9)*.

Complement plays a critical role in several stages of the processing of immune complexes. Complement reacts with immune complexes to inhibit immune precipitation *(10,11)*, solubilize immune aggregates *(12,13)*, and promote immune complex binding to erythrocyte CR1 *(14,15)*. Covalently-incorporated cleavage products of C3 and C4, deposited on immune complexes, following the activation of the complement system, influence the fate of immune complexes by acting as

From: *Methods in Molecular Biology, vol. 150: Complement Methods and Protocols*
Edited by: B. P. Morgan © Humana Press Inc., Totowa, NJ

ligands first for receptors on cells that transport immune complexes through the body and, second, for receptors on cells that take up and process immune complexes.

Model immune complexes have been utilized as tools for exploring the mechanisms of processing and clearance of immune complexes by complement, Fcγ-receptors and the reticuloendothelial system in various animal models including primates and man. Also, immune complexes are employed in the creation of models of disease states thought to be mediated wholly or in part by immune complexes such as serum sickness, the Arthus reaction, and immune complex mediated glomerulonephritis. Alternatively, various methods and techniques have been developed to detect the presence of circulating immune complexes in the serum of individuals with diseases thought to be mediated by immune complexes.

1.1. Generation and Characterization of Model Immune Complexes

Numerous antigens and their corresponding antibodies have been used as model immune complexes. Well characterized immune complexes that have been used in numerous in vivo and in vitro studies include: (1) BSA/anti-BSA *(16)*; (2) Heat-aggregated IgG *(17)*; (3) Tetanus toxoid/antitetanus toxoid *(18)*; (4) HBsAg/anti-HBsAg *(6)*

Aggregated human gamma globulin (AHG), chemically crosslinked immunoglobulin, and immune complexes of defined antigen–antibody composition have all been used as calibrators for immune complex assays by various investigators. Nonspecifically aggregated IgG has been most widely used and a method for producing AHG is described here.

1.2. Assays for Circulating Immune Complexes

Because of the pathogenic role of immune complexes in many diseases, considerable research has been directed towards the development of specific, sensitive, and reliable techniques for their direct demonstration. Many assays for immune complexes have been developed over the last several decades. None should be considered diagnostic for any disorder and relatively few techniques have been put into routine clinical use. The techniques employed for immune complex detection depend on several different characteristics and properties of the immune complex and are listed in **Table 1**.

A few commercial sources of immune complex assay kits are now available, not all of which have been shown to be consistent or reliable. There is an extensive literature, dating from the 1970s and 1980s relating to the utility of immune complex measurement, mainly using solid phase assays, in a variety of diseases *(19–21)*. A detailed discussion of all the methods and techniques for detection of circulating immune complexes is beyond the scope of this chapter. Methods most commonly used are PEG precipitation of immune complexes

Table 1
Techniques for Immune Complex Detection and Characterization

Methods based on physical properties:

Based on changes in solubility	PEG precipitation
	Measurement of cryoglobulins
Based on changes in size	Sucrose density gradient centrifugation
	Gel filtration
	Ultrafiltration
	Electrophoresis and electrofocusing
	Analytical ultracentrifugation

Methods based on biological properties:

Complement-dependent techniques	Microcomplement consumption test
	C1q-dependent techniques
	C1q precipitation in gels
	C1q deviation tests
	C1q-PEG assays
	C1q solid phase binding assays
	C1q fluid phase binding assay
	C3-dependent techniques
	Anti-C3 solid phase assays
	Raji cell assay
	Conglutinin assays
Fc-dependent techniques	Antiglobulin techniques
	RF tests
	Binding to staphylococcal protein A
Cellular techniques	Platelet aggregation test
	EAC lymphocyte rosette formation
	inhibition assay
	Inhibition of antibody-dependent
	cellular cytotoxicity
	Macrophage inhibition assays
	Release of enzymes from eosinophils
	and mast cells

and the detection and estimation of cryoglobulinemia. The one technique of general clinical relevance is the measurement of cryoglobulins in serum from patients. When cryoglobulins are suspected, great care must be taken to avoid loss of immune complexes by cryoprecipitation. Blood and serum must be maintained at or close to 37°C at all times. Because the effects of storage are not clearly defined, samples should be assayed as quickly as possible after separation or stored at –70°C, although there is some evidence that rapid freezing of samples can damage immune complexes *(22)*. Repeat freeze-thaw cycles and heating of the sample to 56°C for inactivation ("decomplementing") are both contraindicated as these procedures introduce the risk of aggregating IgG, leading to false positive results.

1.3. Limitations and Problems of Immune Complex Assays

Each of the various techniques described here and elsewhere for measurement of immune complexes has its own advantages and disadvantages. Some of the assays involve reagents not commercially available, although the reagents needed can be obtained via methods well described elsewhere in this volume. No method differentiates nonspecifically aggregated immunoglobulin from immune complexes and thus great care must be taken in the sample preparation as emphasized in each method. Most tests are also influenced to varying degrees by the concentration of monomeric IgG in serum. A number of substances that may be present in serum are known to interact with C1q and may therefore interfere with the assays based on C1q binding. These substances include single-stranded DNA, lipopolysaccharide and, more importantly, autoantibodies against C1q (this is discussed in greater detail in Chapter 13). In the assays that are cell-based, there is the possibility of interference from the presence of antibodies reacting with cell-surface receptors.

For optimal screening for immune complexes, several assays should be employed that are based on different principles and ideally incorporating known negative, low-positive, and high-positive controls and calibrated with aggregated human IgG.

2. Materials

2.1. Buffers

1. Phosphate buffered saline (PBS) is purchased in tablet form from Oxoid.
2. Complement fixation diluent is purchased in tablet form from Oxoid.
3. 0.9% NaCl is made by dissolving 9 g NaCl in 1 L distilled H_2O.
4. Saturated ammonium sulphate solution is made by dissolving the maximum possible amount of Ammonium sulphate in distilled H_2O at room temperature (20°C).

2.2. Generation of Immune Complexes

1. Water bath at 63°C.
2. Ice bath at 0°C.
3. Human IgG, obtained from commercial sources or isolated from normal serum by DEAE cellulose ion exchange chromatography.

2.3. Immune Complexes from Serum

1. Water bath at 37°C.
2. Gamma counter suitable for Eppendorf microfuge tubes.
3. Refrigerated ultracentrifuge for sucrose density gradient centrifugation.
4. Refrigerated microcentrifuge for pelleting immune complexes.
5. Heated (37°C) centrifuge for separating samples to be tested for cryoglobulins.

3. Methods
3.1. Production of Aggregated Human Gamma Globulin (AHG)

1. Prepare a 6-mg/mL solution of human IgG in 0.9% NaCl.
2. Avoid vigorous mixing, which may cause bubbles to form around the protein particles, thus inhibiting solubilization, which may take several hours.
3. Incubate with occasional stirring at 63°C for 30 min.
4. Immediately place in an ice bath until cool.
5. Centrifuge at 1500g for 15 min, and save the supernatant.
6. Measure the protein concentration in a standard protein assay.
7. Dilute to a concentration of 5.0 mg/mL in CFD, and store at –70°C *(23)*.

3.2. Preparation and Characterization of Immune Complexes from Serum

Various techniques are available to prepare immune complexes from serum and to determine their size and properties of the antigen–antibody interaction such as antibody affinity and the ability of the immune complex to activate complement in the fluid phase or bind to complement receptors. The main methods are outlined below.

3.2.1. Ammonium Sulphate Precipitation (Farr Assay)

The principle underlying this assay is that when ammonium sulphate is added to diluted serum to 50% saturation, most of the immunoglobulin is precipitated, whereas other serum proteins such as albumin, remain in solution. If labeled antigen is premixed with the test serum, only that part of the antigen complexed with antibody is precipitated under these conditions.

1. The antigen is radiolabelled with [125]I (*see* **Note 1**) and allowed to interact with antibody in the serum sample, generally in antigen excess, for 1 h at 37°C so that soluble immune complexes are formed.
2. An equal volume of saturated ammonium sulphate is added to the serum sample and the mixture incubated at ambient temperature for 30 min to precipitate all the immunoglobulin.
3. The precipitates are pelleted by centrifugation, washed with 50% saturated ammonium sulphate solution to remove any free antigen and their radioactive content determined by counting in a gamma counter.

4. The amount of radioactivity in the precipitate is proportional to the amount of antigen bound by the antibody and so results are expressed in terms of the antigen-binding capacity/mL of serum *(24)* (*see* **Note 2**).

3.2.2. Determination of Antibody Affinity

Antibody affinity may be determined by one of several methods. Detailed descriptions and discussion of the relative merits of these methods of determining antibody affinity is beyond the scope of this chapter, however, they are covered in greater detail in *(20)*.

1. Equilibrium Dialysis—this method is generally regarded as the standard (benchmark) method, but is rather cumbersome to perform and consumes large amounts of antigen and antibody. It relies upon the concentration of free hapten equilibrating on both sides of a dialysis membrane, but there is a relatively greater concentration of hapten molecules within the dialysis tubing because of the presence of the corresponding antibody and this is related to the affinity of the antibody for the hapten.
2. Determination of the antibody affinity can be simplified by using a variation of the Farr assay described in **Subheading 3.2.1.** A constant amount of antibody is reacted with increasing concentrations of antigen and left to equilibrate. The concentration of bound antigen is detected in the complex after ammonium sulphate precipitation and the concentration of free antigen determined in the supernatant. If the reciprocal of bound antigen is plotted against the reciprocal of the free antigen concentration and the line extrapolated to $1/[Ag] = 0$, the reciprocal of the bound antigen will equal the reciprocal of the total antibody-combining sites.
3. A major problem with the modification of the Farr assay is the need to wash the precipitates as this introduces the risk of losing some of the precipitate or dissociating some of the complex. A useful modification is the incorporation of $^{22}NaCl$ as a marker of buffer volume. Most of the supernatant can then be removed, without accurate measurement, and the amount of free radioactive antigen remaining in the precipitate estimated from the ^{22}Na counts which will indicate the amount of buffer solution remaining with the precipitate.
4. Functional antibody affinity may be determined by enzyme-linked immunosorbent assay (ELISA). Useful information on relative affinity can be obtained by performing ELISA assays in the presence and absence of diethylamine. This chaotropic agent has a much greater effect in inhibiting the binding of low-, compared with high- affinity antibodies.

3.2.3. Sizing of Immune Complexes by Isopycnic Centrifugation on Sucrose Density Gradients

An estimation of the size of an immune complex can be obtained by centrifugation of the immune complex and free antigen with suitable molecular weight markers on sucrose density gradients.

1. Generate sucrose density gradients by sequentially layering 0.9 mL volumes of 50, 40, 30, 20, and 10% solutions of sucrose in CFD (w/v) into a 5-mL centrifuge tube such that the 50% solution is at the bottom of the tube with subsequent layers of decreasing concentration to the top.
2. Radiolabel with ^{125}I the antigen (*see* **Note 1**) and suitable substances serving as molecular weight markers.
3. Carefully layer onto the tops of the sucrose gradients 100 µL of the relevant immune complex mixture or corresponding antigen or molecular weight marker and centrifuge the tubes at 127,000*g* at 4°C for 3 h.
4. Collect 200 µL fractions sequentially from the bottom of the centrifuge tubes using a long needle and syringe, taking care not to disturb the gradient.
5. Determine the radioactivity in each sample.
6. Plot radioactivity versus fraction number, these will show a shift in the elution profile for the immune complex compared to the profile for free antigen because of the difference in their sizes. The elution profiles for the molecular weight markers are used to estimate the relative molecular weight of the immune complex.

3.2.4. Human Erythrocyte Binding Study to Demonstrate Activation of Human Complement by Immune Complexes

To determine whether the model immune complexes are capable of activating the classical pathway of human complement in the fluid phase an erythrocyte binding assay can be performed. In humans and primates immune complexes bind to erythrocyte CR1 via C3b deposited on the immune complex surface in an immune adherence reaction following activation of the classical complement pathway.

1. Generate immune complexes as described above in which the antigen is radiolabelled with ^{125}I (*see* **Note 1**).
2. Obtain 20 mL of blood from a normal human subject, allow 10 mL to clot and extract the serum.
3. To the remaining 10 mL of blood add 250 µL of 0.5 *M* EDTA, wash the red blood cells three times in PBS at 4°C and resuspend the erythrocytes to a concentration of 1×10^9/mL.
4. Dilute autologous serum 1:5, by volume, in phosphate-buffered saline (PBS); to another similarly diluted serum sample add ethylenediaminetetraacetic acid (EDTA) at a final concentration of 50 m*M* to block complement activity.
5. To 15 µL of the labeled immune complex or antigen only control add 35 µL of diluted autologous serum, or serum/EDTA as a negative control.
6. Incubate at 37°C for 15 min for complement activation to occur and to allow deposition of C3b on any immune complexes that have formed in the antigen–antibody mixture.
7. Add 200 µL of autologous erythrocytes at a concentration of 1×10^9/mL and incubate at 37°C for 5 min.

8. Layer duplicate 75 µL samples of treated erythrocytes onto 150 µL of phthalate oil in Eppendorf tubes microcentrifuge the tubes (12,000g, 5 min).
9. Snap-freeze the tubes at –70°C and then cut through the oil layer with a scalpel blade, enabling the cell pellet and the supernatant fractions to be counted separately in a gamma counter.
10. From these counts the percentage of total radioactivity, and therefore, immune complexes bound to the erythrocyte can be determined (see **Note 3**).

3.3. Complement Mediated Inhibition of Immune Precipitation and Solubilization of Immune Complexes

These assays measure the capacity of complement to facilitate the precipitation of soluble immune complexes and the incorporation of complement components into the aggregates *(25)*. They may also be used to measure complement mediated solubilization of immune aggregates, a function mediated by the alternative pathway and enhanced by the classical pathway *(12)*.

3.3.1. Solubilization Assay

Solubilization of an immune precipitate by serum provides the basis of a simple quantatative assessment of the integrity of complement function. An assay to test the solubilization capacity of human serum utilizing preformed radiolabeled precipitates of BSA/anti-BSA is outlined below.

1. Radiolabel BSA with ^{125}I (see **Note 1**).
2. Heat-inactivate polyclonal rabbit anti-BSA antiserum by incubation at 56°C for 30 min and clear aggregates by centrifugation at 40,000g for 4 h.
3. Incubate antigen and antibody at 37°C for 1 h and then overnight at 4°C to allow immune precipitation to occur.
4. Wash the BSA/rabbit anti-BSA immune precipitate six times in cold PBS and store at –70°C until needed.
5. Add 5 µL of ^{125}I BSA/anti-BSA immune precipitate suspension containing 1 µg of antibody to 50 µL of test serum diluted 1:3 in CFD.
6. After incubation at 37°C for 60 min, add 1.5 mL of ice-cold PBS to stop any further complement activation.
7. Centrifuge the mixture at 1500g for 10 min at 4°C and discard the supernatant
8. The percentage of solubilization is calculated as:

$$(\text{total cpm} - \text{pellet cpm/total cpm}) \times 100$$

9. To calculate the complement-mediated solubilization, subtract the spontaneous solubilization occurring in complement-depleted (heat-inactivated) serum from the total solubilization. In this assay, spontaneous solubilization is usually around 5–10% *(11)*.

3.3.2. Assay for Inhibition of Immune Precipitation

The following is a simple method for determining the degree of inhibition of immune precipitation by complement in serum.

1. In each of a series of tubes, preincubate a radiolabeled antigen, e.g., 1 µg of ^{125}I BSA (*see* **Note 1**), in 100 µL of human serum at 37°C for 5 min. An identical series is set up in which heat-treated serum is substituted (control).
2. Add the corresponding antibody, e.g., 2 µL of heat inactivated rabbit anti-BSA antiserum, to each tube, quickly mix, and incubate at 37°C.
3. Remove pairs of tubes (complement and control) at various time points up to 60 min, add 1 mL of ice-cold PBS and store on ice till end of timecourse.
4. Centrifuge the tubes at 3000g for 10 min at 4°C.
5. Count the radioactivity in the pellets and supernatants and plot the proportion of offered counts in the supernatant at each timepoint for complement and control series.
6. The difference between complement and control at each time point (greater proportion of radioactivity in supernatant in presence of complement) reflects the degree of inhibition of immune precipitation by complement (*10*).

3.4. Assays for Circulating Immune Complexes

3.4.1. Polyethylene Glycol (PEG) Precipitation of Immune Complexes

One of the simplest methods available for detection of immune complexes is their precipitation from serum samples by the addition of polyethylene glycol (PEG) (*26,27*). This procedure is suitable for nonspecialist laboratories, but should only be considered as a nonspecific screening test for the presence of immune complexes (*see* **Note 4**).

1. Mix 100 µL of test serum with an equal volume of 4% PEG 4000 (w/v) in PBS in a precipitin tube.
2. Incubate for 1 h at 4°C.
3. Centrifuge at 1000g for 1 h at 4°C.
4. Remove clear supernatant and wash the pellet twice with 2% PEG in PBS.
5. Dissolve the PEG precipitate in PBS and measure the protein content by standard protein assays.
6. Compare the amount of precipitation ((g/mL test serum) with that obtained in normal human serum studied in parallel.

3.4.2. Determination of Cryoglobulinaemia

1. Collect blood from the patient, taking care at all stages of sampling and transportation to the laboratory to maintain the blood close to 37°C. Ideally, transport in a vacuum flask prefilled with water from a 37°C water bath.
2. Leave the blood sample to clot overnight at 37°C.

3. Centrifuge the sample in a heated centrifuge at 37°C to separate the serum.
4. Place the serum sample at 4°C for 5 d for the cryoprecipitate to form.
5. Centrifuge the sample at 1000g at 4°C for 15 min in a graduated tube; the cryo-precipitate is measured as a percentage by volume of the serum sample.
6. The degree of cryoglobulinemia is expressed as this cryoprecipitate percentage by volume (*see* **Note 5**).

4. Notes

1. Radiolabeling of proteins with ^{125}I is best performed using the iodogen method. Iodogen is prepared as a stock solution at 0.4 mg/mL in dichloromethane and 40 µL aliquots placed in 2 mL round-bottomed glass tubes. The tubes are placed in a fume hood for several hours to allow the solvent to evaporate. Dissolve or dilute protein sample in 50 m*M* borate buffer pH 8.3 containing 75 m*M* NaCl. Add 0.1–1 mCi Na^{125}I (IMS30; Amersham Pharmacia). Incubate on ice in fume hood for 5 min. Separate bound from free radioactivity by gel filtration on a column of Sepharose G25 (PD10; Amersham Pharmacia). Collect 0.5 mL fractions off the column, measure radioactivity and pool peak of labelled protein. Store at 4°C with NaN$_3$. Use within two weeks of labeling.
2. For accurate determinations in research procedures, the results need to be calculated to the same degree of antigen excess for each antiserum. To achieve this, constant amounts of antigen are added to a series of dilutions of serum and the antigen-binding capacity determined at each dilution. The convention is to calculate the results at the 33% end point, i.e., the dilution of serum which binds 33% of the added antigen *(24)*.
3. The hemolytic complement assays described in Chapter 4 can also be employed to determine the relative abilities of immune complexes to activate the classical and alternative pathways of complement in the fluid phase. The serum, to be employed in the hemolytic assays, is incubated with an appropriate amount of immune complex at 37°C for 15 min before being utilized in the assay alongside serum not incubated with immune complexes. The reduction in the hemolytic complement activity of the serum samples preincubation with immune complexes, compared to untreated control serum, is directly related to the degree of complement activation and consumption by those immune complexes.
4. PEG is an uncharged linear polymer. The addition of PEG to serum causes a precipitation of proteins proportional to the concentration of PEG *(26)*. The degree of precipitation of each protein is proportional to its molecular weight and thus at low concentrations of PEG, only high-molecular weight proteins and immune complexes are precipitated. An abnormal increase in the protein concentration of the PEG precipitate from a serum sample correlates with the level of circulating immune complexes. This measurement will be affected by the level of various serum proteins and therefore the result obtained is not reliable for an accurate quantitative measurement of the concentration of circulating immune complexes.

5. Alternatively, the cryoprecipitate can be washed in ice-cold buffer, dissolved in buffer at 37°C and the protein content measured in a standard protein assay. This enables the amount of cryoglobulins present to be accurately quantified.

References

1. von Pirquet, C. F., and Schick, B. (1905) Die Serum Krankeit. (F. Deuticke, ed.), Leipzig, Germany.
2. Arthus, M. and Breton, M. (1903) Lesions cutanees produties par les injections de serum de cheval chez le lapin anaphylactise par et pour ce serum. *C. R. Soc. Biol.* **55,** 817–820.
3. Dixon, F. J. (1963) The role of antigen-antibody complexes in disease. *Harvey Lect.* **58,** 21–52.
4. Germuth, F. G. and Rodriguez (1973) *Immune Complex Glomerulonephritis,* 1st ed., Little, Brown & Co., Boston.
5. Atkinson, J. P. (1986) Complement activation and complement receptors in systemic lupus erythematosus. *Springer Sem. Immunopathol.* **9,** 179–194.
6. Davies, K. A., Peters, A. M., Beynon, H. L. C., and Walport, M. J. (1992) Immune complex processing in patients with systemic lupus erythematosus-in vivo imaging and clearance studies. *J. Clin. Invest.* **90,** 2075–2083.
7. Klaus, G. G. B and Humphrey, J. H. (1977) The generation of memory cells. I. The role of C3 in the generation of B memory cells. *Immunol.* **33,** 31–40.
8. Klaus, G. G. B. 1978. The generation of memory cells. II. Generation of, B. memory cells with preformed antigen-antibody complexes. *Immunol.* **34,** 643–652.
9. Embling, P. H., Evans, H. Guttierez, C. Holborow, E. J., Johns, P. Johnson, P. M., et al. (1978) The structural requirements for immunoglobulin aggregates to localise in germinal centres. *Immunol.* **34,** 781–786.
10. Schifferli, J. A., Bartolotti, S. R., and Peters, D. K. (1980) Inhibition of immune precipitation by complement. *Clin. Exp. Immunol.* **42,** 387–394.
11. Schifferli, J. A., Morris, S. M., Dash, A. and Peters, D. K. (1981) Complement-mediated solubilisation in patients with systemic lupus erythematosus, nephritis or vasculitis. *Clin. Exp. Immunol.* **46,** 557–564.
12. Miller, G. W. and Nussenzweig, V. (1975) A new complement function: solubilisation of antigen-antibody aggregates. *Proc. Natl. Acad. Sci. USA* **72,** 418–422.
13. Takahashi, M., Tack, B. F., and Nussenzweig, V. (1977) Requirements for the solubilisation of immune aggregates by complement: assembly of a factor B-dependent C3-convertase on the immune complexes. *J. Exp. Med.* **145,** 86–100.
14. Walport, M. J. and Lachmann, P. J. (1988) Erythrocyte complement receptor type 1, immune complexes, and the rheumatic diseases. *Arthritis Rheum.* **31,** 153–158.
15. Liszewski, M. K. and Atkinson, J. P. (1991) The role of complement in autoimmunity, in *Systemic Autoimmunity* (Reichlin, M. R. and Bigazzi, B., eds.), Marcel Dekker, New York, pp. 13–37.
16. Cornacoff, J. B., Hebert, L. A., Smead, W. L., Vanaman, M. E., Birmingham, D. J., and Waxman, F. J. (1983) Primate erythrocyte-immune complex clearing mechanism. *J. Clin. Invest.* **71,** 236–247.

17. Lobotto, S. Daha, M. R., Voetman, A. A., Evers-Schouten, J. H., van Es, A. A., Pauwels, E. K., and van Es, L. A. (1987) Clearance of soluble aggregates of immunoglobulin G in healthy volunteers and chimpanzees. *Clin. Exp. Immunol.* **69,** 133–141.

18. Schifferli, J. A., N. g., Y. C., Estreicher, J., and Walport, M. J. (1988) The clearance of tetanus toxoid/anti-teanus toxoid immune complexes from the circulation of humans. Complement- and erythrocyte complement receptor 1-dependent mechanisms. J. Immunol. 140, 899–904.   Au: explain Should 'G."? name

19. Abrass, C. K., Nies, K. M., Louie, J. S., Border, W. A., and Glassock, R. J. (1980) Correlation and predictive accuracy of circulating immune complexes with disease activity in patients with systemic lupus erythematosus. *Arthritis Rheum.* **23,** 273–282.

20. Davies, K. A. (1992) Immune complexes and disease. *European J. Intern. Med.* **3,** 95–108.

21. Levinson, S. S. and Goldman, J. O. (1987) Evaluation of anti-C1q capture assay for detecting circulating immune complexes and comparison with polyethylene glycolimmunoglobulin G, C1q-binding, and Raji cell methods. *J. Clin. Microbiol.* **25,** 1567–1569.

22. WHO (1977) The role of immune complexes in disease. *WHO Tech. Rep.* **Ser. :606.**

23. Zubler, R. Lange, G. Lambert, P., and Miescher, P. (1976) Detection of immune complexes in unheated sera by a modified ^{125}I C1q binding test. *J. Immunol.* **116,** 232.

24. Hudson, L. and Hay, F. C., eds., *Practical Immunology,* 3rd ed., Blackwell Scientific, Oxford, ch. 6.

25. Weigle, W. O. and Maurer, P. H. (1957) The effect of complement on soluble antigen–antibody complexes. *J. Immunol.* **79,** 211.

26. Creighton, W. D., Lambert, P. H., and Miescher, P. A. (1973) Detection of antibodies and soluble antigen-antibody complexes by precipitation with polyethylene glycol. *J. Immunol.* **111,** 1219.

27. Chia, D., Barnett, R. V., Yamagata, J. Knutson, D. Restivo, C., and Furst, D. (1979) Quantitation and characterisation of soluble immune complexes precipitated from sera by polyethylene glycol (PEG). *Clin. Exp. Immunol.* **37,** 399.

16

Knocking Out Complement Genes

Anne E. Bygrave and Marina Botto

1. Introduction

The modification of chromosomal loci by homologous recombination is commonly referred to as gene targeting, the importance of which needs no emphasis. Gene targeting has been widely used in mice to make a variety of mutations in many different loci, including those of the complement components, so that the phenotype of genetic modifications may be assessed in the entire organism. Gene targeting was first achieved in mammalian cells in 1985 *(1,2)*. With the isolation of embryonic stem (ES) cells from the inner cell mass of early mouse embryos *(3,4)* and from the realization that these cells could then be cultured for a substantial period of time and still contribute to the germline of a mouse after reintroduction into a blastocyst, even after genetic modification *(5,6)*, a powerful new method for analyzing gene function in vivo has emerged.

Gene-targeting experiments can be divided into three basic parts. First, the design and construction of a targeting vector. Second, the transfection of vector DNA into ES cells followed by the selection and screening for homologous recombinants. Third, the generation and breeding of chimeric mice to obtain germline transmission of the mutant allele. The following sections will cover some of the practical aspects of gene targeting with special emphasis on the ES culture conditions.

1.1. Targeting Vectors

A targeting vector is designed to recombine with and mutate a specific chromosomal locus. In this section, we will outline some general principles about the design of the targeting vectors that should be kept in mind. Technical aspects in the construction are beyond the limits of this chapter. Two distinct

From: *Methods in Molecular Biology, vol. 150: Complement Methods and Protocols*
Edited by: B. P. Morgan © Humana Press Inc., Totowa, NJ

vector designs are commonly used for targeting mammalian cells: replacement and insertion vectors. To date, all complement knockout mice have been generated using replacement vectors. A typical replacement vector consists of two regions of homology to the target locus flanking a positive selection marker, bacterial plasmid sequences and a unique restriction enzyme site outside the region of homology (for linearization of the targeting vector prior to transfection). Generally, the positive selection marker is inserted into, or replaces, an exon of the target gene (*see* **Fig. 1** for the basic structure of a replacement vector and its method of recombination). Addition of a negative selection marker *(7)* to the replacement vector was an improvement which allowed enrichment against random integration events. The final product of homologous recombination with a replacement vector is replacement—by the positive selection marker—of the endogenous sequence found between the two regions of homology. Generally, such a targeting event will cause a gross alteration in the structure of a locus, however, more elaborate targeting systems have been devised for more subtle alterations such as the introduction of a single nucleotide substitution into a locus *(8,9)*.

The targeting vector construction is the central feature of any gene-targeting experiment. When designing a targeting vector, it is essential to consider all the possible strategies available in the literature and bear in mind all the points listed below. Vector design is fundamentally limited by the type of mutation that is required and by the method of selection and screening for the occurrence of homologous recombination. Certain mutations in a gene may not be adequate for generating a genuine null allele, that is an allele which produces no mRNA and hence no protein product. Production of truncated or modified protein products from an incorrectly targeted gene may result in a normal or even abnormal biological activity. For this reason, it is necessary to confirm, after generating mice carrying targeted alleles, that the allele is null by carrying out RNA and protein analyses. If the intended end product of a gene-targeting event is to ablate a specific known function of a gene, then a functional assay of the protein biological activity is desirable.

The following is a list of basic considerations for the design of a replacement vector:

1. insert the positive selection marker into an upstream exon;
2. delete 5' exons preferentially, or the entire gene if possible;
3. all of the exons to be interrupted should not contain a unit number of codons so that a product with some function is not produced after removal of the mutated exons by RNA splicing;
4. if particular exons are believed to encode a part of a protein essential for a function to be ablated then these exons can be targeted preferentially;

A Homologous Recombination:

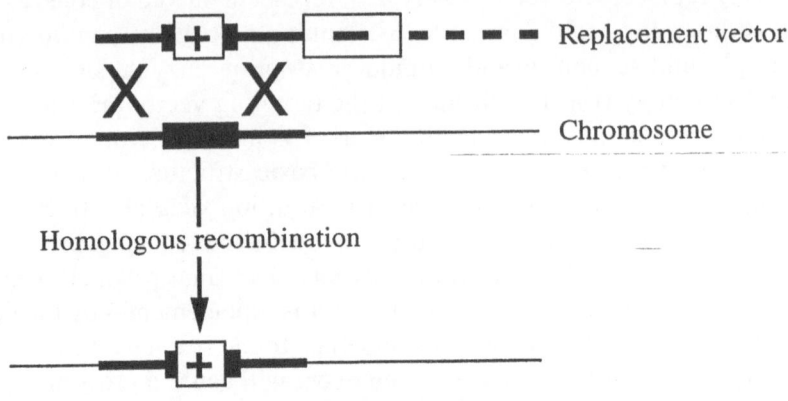

Homologous recombination

resistant to positive and negative selection

B Random Recombination:

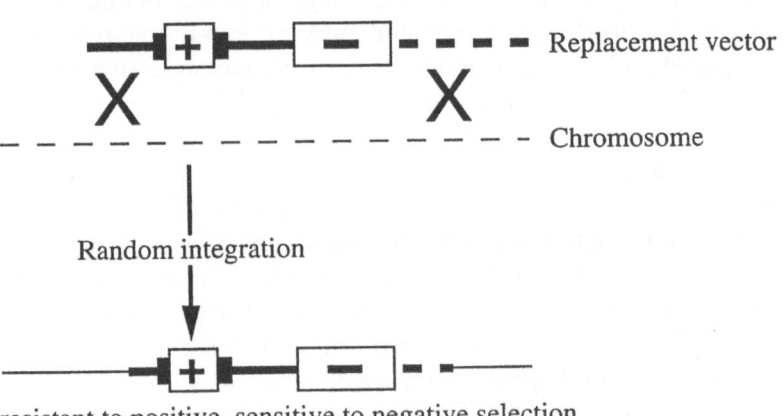

Random integration

resistant to positive, sensitive to negative selection

Fig. 1. Schematic representation of gene targeting by homologous recombination using a conventional replacement vector. a) Positive/negative selection medium enriches for the rare homologous recombination events where the positive selection marker (+) is retained and the negative selection marker (−) is lost. The positive selection marker disrupts the endogenous target sequence (black box). b) The conventional replacement vector usually integrates randomly into the genome. When this occurs the negative selection marker remains intact and the cell is killed by negative selection medium.

5. it is important to note that any gene to be targeted may be in the close proximity of another gene which could be influenced by the introduced changes. In these situations, a phenotype could be because of a multiple gene disruption rather than just the disruption of the targeted gene.

A targeting vector should be constructed using genomic DNA from the same strain of mouse to that from which the ES cells were isolated. This is because isogenic DNA has been shown to increase targeting frequencies by as much as 20-fold when used in targeting vectors *(10)*. The homologous flanking regions should total ideally between 5 and 8 kb. Several groups have shown a correlation between the length of homology and the efficiency of targeting a locus *(5,11)*. A generally adopted principal is that the larger the homology, the higher the recombination frequency. A physical limit is imposed on the size of the vectors used by the difficulty with which larger fragments of DNA can be safely manipulated in vitro, and by the need to retain a unique restriction site outside the regions of homology.

1.2. Embryo-Derived Stem (ES) Cells

Murine ES cell lines are derived from the inner cell mass of preimplantation mouse embryos. This inner cell mass is composed of a small group of pluripotent cells that can develop to form the entire embryo. The precise procedure for culturing ES cells varies from laboratory to laboratory, but there are two major requirements for ES cell culture in the context of gene targeting projects: (1) the cells must be maintained in their undifferentiated state and (2) the cells must retain a normal karyotype, which is a prerequisite for germline transmission.

The most successful lines for gene knockout technology so far have been derived from the 129 mouse substrains. Most ES cells are male as these are more stable in culture. 129ES cells grow rapidly and on injection into early mouse embryos are able to integrate successfully, producing chimeric mice carrying ES cell-derived germ cells. This is essentially the basis of the knockout technique. Successful culture for the use of ES cells in the formation of chimeras depends on keeping the cells in the undifferentiated state. The correct culture medium, culture conditions, and treatment by the tissue culturist is therefore essential. A useful piece of advice is to continue culturing ES cells in the same way they were established; do not change anything!

Most ES cells grow best on another layer of cells that mimic in some way the conditions in vivo. These cells are known as feeder layers. The two best-known feeder cells for ES culture are STO fibroblasts (thioguanine and oubain resistant subline of SIM mouse fibroblast) and primary embryonic fibroblasts (PEF). STO fibroblasts can be used up to a passage number of approximately 22 before they become senile. They are neomycin resistant, and some lines (e.g., SNL) also produce leukemia inhibitory factor (LIF), a growth factor essential for ES cells that helps to prevent differentiation. Some ES lines do not grow well on STO or SNL fibroblasts. In these cases, primary embryonic fibroblasts prepared from neomycin resistant mouse embryos should be used as feeder layers. Both STO and PEF cells need to be treated to stop the cells from dividing before they can be used as feeders. Irradiation by a suitable source with a dosage that prevents cell division, or treatment with the antimitotic drug Mitomycin C are both methods that have been successfully used. Some ES cells have been developed that do not need a feeder layer (such as E14). There are advantages to this type of cell since the work preparing feeder layers is eliminated and also the cells can be transfected with gene-targeting vectors containing not only the neomycin resistance gene, but also other selection markers.

2. Materials
2.1. Reagents for Culturing ES Cells

All reagents for tissue culture are from Life Technologies unless otherwise stated.

1. Dulbecco's phosphate-buffered saline without Ca^{2+} and Mg^{2+} (PBSA).
2. Dulbecco's modified eagles medium (DMEM), 4.5 g/L glucose + Na_2HCO_3 and store at 4°C.
3. Penicillin/Streptomycin: stock solution is 100×, kept at −20°C.
4. L-Glutamine: stock solution is 100×, kept at −20°C.
5. MEM Nonessential amino acids: stock solution is 100×, kept at 4°C.
6. ES medium contains the following: DMEM high glucose; 15% ES tested fetal calf serum (FCS); 1000 U/mL LIF; 0.1 mM Nonessential amino acids; 2 mM L-glutamine; Penicillin and streptomycin (final concentrations of 50 µg/mL each); 0.1 mM β-mercaptoethanol.
7. Murine LIF from GibcoBRL as "ESGRO."
8. β-mercaptoethanol 1000x stock solution: 10^{-1} M β-mercaptoethanol in PBSA, filter sterilized, stored at 4°C for not more than 2–3 wk.
9. FCS tested for toxicity and cloning efficiency on ES cells.
10. Gelatin solution: 0.1% solution of tissue culture grade gelatin dissolved by autoclaving in tissue culture grade water. Store at room temperature.
11. Mitomycin C (Sigma). Dissolve 2 mg in 5 mL TC grade water or culture medium. Add to 200 mL of culture medium to give a working solution of 10 µg/mL. Store at 4°C for no more than a week and then discard (*see* **Note 1**).
12. Freezing medium (2×): 80% FCS, 20% Dimethyl sulphoxide (DMSO). Store at 4°C. Small aliquots can also be frozen at −20°C if required.
13. Trypsin/Versene solution. This can be frozen at −20°C in aliquots of 5–10 mL. Defrost an aliquot before use and keep whatever is left at 4°C for short periods.

14. Neomycin: also known as geneticin sulphate (or G418). Dissolve in PBS and filter sterilize. Store at 4°C; long-term aliquots at –20°C.
15. Colcemid (from Sigma). Store a stock solution of 1000× (20 µg/mL) in DMEM at –20°C. Make up a working solution of 0.02 µg/mL and store at 4°C until use.
16. Gancyclovir (Cytovene, Syntex). Dissolve in PBSA and filter sterilize. Stock solution should be 2 mM. Store at –20°C in small aliquots. Add to culture medium just before use at a working strength of 2 µM.
17. Sterile tissue culture plasticware.

3. Methods

3.1. Preparation of Gelatin-Coated Tissue Culture Plates

ES cells are grown on gelatin coated dishes (with feeder cell layers). The protocol is the following.

1. Add gelatin solution to tissue culture dishes—enough to cover the base of the dish.
2. Leave for a minimum of 1 h at room temp. The plates can also be put into a 37°C incubator for several days until needed.
3. Immediately before use, aspirate the gelatin solution to waste and add the cell solution directly to the dish.
4. Never allow the gelatin-coated dishes to dry out or patchy and uneven cell growth will result, sometimes in concentric circles following the evaporation path.

3.2. Making Primary Embryonic Fibroblasts for Feeder Layers

1. Kill pregnant female mouse (neomycin resistant) at day 13 or 14 of gestation
2. Do all dissections in a sterile tissue culture hood.
3. Dip mouse in 70% ethanol. Peg out on board.
4. Open outer skin leaving body cavity intact. Peg skin away from body. Use fresh ethanol dipped instruments to open body cavity.
5. Peg out body wall away from body. Dip instruments again.
6. Dissect out uterus and place into a 90-mm Petri dish.
7. Dissect out embryos and remove extra-embryonic membranes and placenta.
8. Dip embryos in sterile PBSA. Dissect away soft organs and viscera (brain, liver, gut, and chest organs).
9. Put remaining carcass into sterile tissue culture grade trypsin solution (e.g., 0.25% trypsin plus ethylenediaminetetraacetic acid (EDTA) in PBSA).
10. Mince very finely.
11. Transfer all embryos to a 50-mL tube and make up to 2 mL/embryo with trypsin. Put at 37°C for about 30 min. Pipet up and down every 10 min.
12. Add an equal volume of medium (DMEM + 10% FCS). Mix and let larger lumps settle out for 1–2 min.
13. Remove supernatant and plate out.
14. Plate out one embryo per 175 cm^2 flask in 50 mL of medium. Incubate at 37°C, 95% O_2, 5% CO_2.

15. Change medium the next day.
16. Grow to confluence.
17. Freeze in freezing medium all flasks at one embryo (equivalent to one flask) per vial except for one.
18. Split remaining flask 1/5 into 5×175 cm^2 flasks.
19. Grow to confluence. Split 1/5 again into 25 flasks.
20. Grow to 80–90% confluence.
21. Treat with Mitomycin C to inhibit cell division. Trypsinize and count cells using a hemocytometer. Freeze cells at 1×10^6 cells per ampule, or multiples of this density as required.

3.3. Preparation of Mitotically Inactive Feeder Layers from Fibroblast Lines

1. Fibroblast lines are grown in DMEM + 10% FCS
2. Grow up a sufficient number of cells for treatment (10–20 × 90 mm dishes for STO/SNL or 5–25 × 175 cm^2 flasks for PEF).
3. When the cells are about 80–90% confluent, i.e., nearly, but not quite covering the whole of the dish surface, change the medium to stimulate cell division.
4. Two hours later change the medium for medium containing Mitomycin C (final concentration 10 μg/mL). Run the solution gently down the side of the plate or drip gently over the surface of the cells so as not to dislodge them since they are loosely attached at this stage.
5. Culture at 37°C for a further 3–4 h for STO/SNL cells or 1.5–2 h for PEF cells.
6. Aspirate the Mitomycin C medium to waste into chloros to deactivate it.
7. Wash the cells very gently twice with 10 mL of PBSA.
8. Add required amount of trypsin and rock the dish or flask immediately to spread the trypsin over the surface of the cells. Rock until the cells come off (very fast at room temperature). Immediately, add a minimum of a 3× volume of fresh culture medium to each and pipet the cells gently to break up clumps.
9. Spin down at 800*g* for 10 min, in aliquots of 15 mL or less.
10. Aspirate all the medium from the pellets. Flick pellets firmly to loosen cells then resuspend in fresh culture medium.
11. Pool all the cells together into 50-mL tubes.
12. Count a sample on a hemocytometer. Dilute the cells to a convenient density e.g., $1–1.4 \times 10^6$/mL.
13. Either freeze down or plate out on gelatin-coated dishes at the following densities: STO/SNL cells: 0.3×10^6/35 mm tissue culture dish; 1×10^6/60 mm tissue culture dish; 3×10^6/90 mm tissue culture dish; 3×10^6/multiwell plate; PEF cells: 0.3×10^6/35 mm tissue culture dish; 0.8×10^6/60 mm tissue culture dish; 1.4×10^6/90 mm tissue culture dish; 1.5×10^6/multiwell plate (*see* **Note 2**).

3.4. Passage and Maintenance of ES Cells

1. Grow ES cells in ES medium; change medium daily; passage cells every 2 d or whenever confluent by trypsinization (*see* **Note 3**).

2. For trypsinization, aspirate medium from the cells and wash monolayer gently in PBSA. Add warmed trypsin solution (sufficient to just cover the cells). Incubate at RT for 1–2 min knocking the plate or flask occasionally to loosen the cells.
3. Make sure all the cells have come off the bottom of the dish before the next step.
4. Add medium to the cell suspension (twice as much medium as trypsin used) and mix well.
5. Triturate the cell suspension to break up cell clumps by passage through a plugged Pasteur pipet, a suitable graduated pipet, or with very small volumes, a 1-mL micropipet Gilson with a plugged sterile tip.
6. Check the cells under the microscope. If not in a single-cell suspension, triturate more and check again.
7. Dilute 1:3 with fresh, prewarmed ES medium and mix again before plating out onto feeder plates. Do not dilute the cells any more or you may increase the risk of favoring abnormal or differentiating cells (*see* **Notes 4** and **5**).

3.5. Freezing of ES Cells

1. Grow ES cells until near confluence before freezing (see **Note 3**).
2. Change the medium on the cells 2 h before freezing to stimulate division.
3. Aspirate the medium on the cells to waste and wash twice with PBSA.
4. Trypsinize as described in **Subheading 3.4.** Count cells if necessary. *See* **Note 6**.
5. Spin down the washed, trypsinized ES cells at 800*g* for 5 min.
6. Aspirate supernatent to waste.
7. Loosen pellet by firm flicking and resuspend cells in culture medium in half the required volume.
8. Place 0.5 mL cells into labeled cryotubes, add 0.5 mL of 2× freezing medium slowly with constant mixing to obtain a final total volume of 1 mL per cryotube.
9. Place ampules into a slow freezing container (*see* **Note 7**), transfer container to a –70°C freezer and leave overnight.
10. The following day, transfer ampules into liquid nitrogen for long-term storage.

3.6. Thawing ES Cells

1. Remove an ampoule from liquid nitrogen storage and agitate in a 37°C water bath until just defrosted (1–2 min).
2. Transfer to a conical tissue culture tube.
3. Add 5 mL of warmed ES medium dropwise to the cells with constant gentle mixing (tapping the tube with the fingers to swirl the medium).
4. Spin down at 800*g* for 5 min at room temperature.
5. Aspirate medium to waste.
6. Loosen the pellet in the residual medium remaining in the bottom of the tube by firm flicking with the fingers.
7. Add a few drops of medium and mix again.
8. Make up to the required volume with warmed medium and plate out onto feeder layers (*see* **Note 8**).

3.7. Production of Targeted ES Cell Clones

There are several different methods available to introduce the DNA into ES cells *(12)*. Electroporation is technically the easiest and has recently become the most popular. An electric shock is passed across the cells allowing the DNA molecules to enter through temporary pores in the cell membranes. This procedure, however, results in the death of 50% of the cells even under optimal conditions. The parameters that can influence efficiency and cell viability are voltage, ion concentration, DNA concentration, and cell concentration; these should be tested for each ES cell line used. The following parameters are based on our experience.

1. Resuspend 1×10^7 ES cells in 800 µL of PBSA.
2. Add between 10 and 20 µg of digested construct DNA in 50 µL of PBSA.
3. Place in 0.4-cm electrode gap electroporation cuvets (BioRad).
4. Deliver an electric pulse across each cuvet using a BioRad gene pulser preset for 240 V, 500 µF.
5. Stand cells on ice for 10–20 min, then replate in 3×90 mm tissue culture dishes.
6. At 24–48 h posttransfection change the ES cell medium for selective medium [if the construct was designed for positive and negative selection with both the neomycin resistance gene and the thymidine kinase gene the medium should be supplemented with G418 (250–450 µg/mL depending on cell line) and Gancyclovir (2 µ*M*)].
7. After about 8 d of selection, drug resistant colonies should appear and be ready for screening by polymerase chain reaction (PCR) or Southern blot analysis (*see* **Note 9**).

3.8. Karyotypic Analysis

1. Plate out cells from recombinant clones to give an approximate 75% confluency on day of harvest.
2. Change the medium on day of harvest.
3. 2 h later, add Colcemid solution to a final concentration of 0.02 µg/mL.
4. Incubate for exactly 1 h, then wash cells gently with PBSA. Trypsinize cells off plate (*see* **Subheading 3.4.**), quench with medium, and spin down cell pellet.
5. Resuspend pellet again with PBSA and spin down.
6. Resuspend pellet gently in hypotonic solution (0.56% w/v KCl), to obtain a single-cell suspension. Stand at RT for exactly 6 min.
7. Spin down at 500*g* for 5 min.
8. Aspirate hypotonic solution to waste, leaving a drop in which to resuspend the pellet.
9. Flicking the tube constantly, add 5 mL of ice-cold freshly made fixative, dropwise. (Fixative: 3:1 mixture of methanol to glacial acetic acid. Make up fresh, cool, and keep on ice during use).
10. Stand at RT for 5 min and spin down cells as above.

11. Aspirate the fixative to waste, leaving 1 mL of fixative in the tube to resuspend the loose pellet. (The cells are very loosely packed at this stage. It is important not to aspirate all the fixative from the tube or the cells will be aspirated to waste also.)
12. Repeat fixing procedure a further three times.
13. Leave the final cell pellet in a volume of 0.5–1 mL.
14. Taking some of the cell suspension into a Pasteur pipet fitted with a rubber teat, allow single drops of cells to fall from a height of between 15 to 20 cm onto dry acid-ethanol washed slides (*see* **Note 10**) under a constant flow of air (e.g., in a sterile hood or fume hood.). The drops should not fall on top of one another. Discard the slide if this happens. Two drops will comfortably fit on one slide, one at each end.
15. Stain the slides by immersion for 20 min in Geimsa 1:20 in Gurr's buffer. Rinse excess stain off gently under running tap water. Air-dry and observe under oil immersion with a 100× oil objective. Count the number of chromosomes in 40 cells per clone and calculate the percentage euploidy (*see* **Note 11**).

3.9. Production and Analysis of Chimeric Mice from Targeted ES Cells

3.9.1 Blastocyst Injection

1. Harvest the recombinant ES cells to be injected into a single-cell suspension and place in a small Petri dish.
2. View under an inverted microscope and collect 15–20 cells into a finely drawn microscopic glass pipet.
3. Using a microinjection apparatus and while viewing under the microscope, inject the cells through the zona pellucida, via a cell–cell junction, into the blastocoele cavity (*see* **Note 12**).
4. Culture in ES medium for 1–3 h at 37°C in 95% O_2, 5% CO_2) to allow the blastocyst to re-expand, and the blastocoele to reform.
5. Transfer the blastocysts to the uterus of a 2.5-d postcoitum pseudopregnant female mouse so that they may continue development in vivo. Generally, 6–10 blastocysts are transferred into a single uterine horn. Gestation takes approximately another 15.5 d (total gestation 19 d). Pseudopregnant females are obtained by mating females with vasectomized males. For further technical details, *see* **refs. *13* and *14***.

3.9.2. Use of Coat Color to Identify Chimerism and Germ Cell Transmission

Pigmentation markers can most easily be used for the assay of germline transmission.

1. ES cells have been obtained from the inner cell mass of agouti mice (commonly the 129/Sv strain) that have hairs with black, yellow, and white stripes, giving the coat an overall midbrown cast. The belly is a pale sandy color. The host blastocysts are

taken from C57BL/6 donors that have a black (nonagouti) coat color. Chimeric pups formed from agouti ES cells that combined with the black donor blastocyst cells are thus identified by the presence of agouti color on a black background.

2. Those pups with the highest percentage of agouti in the coat are the most chimeric and most useful. With some ES cell clones, chimeras with near to 100% agouti coat color can be obtained. Only male chimeric mice are bred.

3. Male chimeras can have both host derived and ES cell-derived testes and sperm cells. Female embryos may be converted into male embryos by the integration of the XY ES cells. In these sex-changed chimeras, all of the spermatocytes, and hence all their sperm, will be derived from the ES cell line, so giving a higher chance of germline transmission.

4. A good pointer for a successful cell clone is to find a higher proportion of male pups in the litters resulting from the embryo transfers. This is known as sex ratio distortion. Hermaphrodites may also sometimes be seen.

5. All high percentage agouti male chimeras are mated with black wild-type females (C57BL/6). For this reason, the vast majority of gene-targeted mice are initially produced on a mixed genetic background ($129 \times$ C57BL/6) that can influence the phenotypic analysis. The agouti coat color gene is dominant over the black (nonagouti) gene. Agouti sperm fertilizing black ova will result in agouti pups.

6. Fifty percent of the agouti pups should contain the knockout allele. Black pups are discarded as wild type (*see* **Note 13**).

7. A small piece of tail (approx 1–2 mm) is snipped from each agouti pup and DNA is extracted and screened for the mutant allele using the protocol developed for screening recombinant ES clones (*see* **Note 9**).

8. The mice carrying the gene-targeted allele are heterozygous and are mated together to produce progeny homozygous for the genetic alteration. Subsequent breeding of the germline transmitting chimeras with 129 females is a useful strategy for generating knockouts in a pure inbred genetic background in just two generations (*see* **Note 14**).

4. Notes

1. Mitomycin C is toxic by inhalation or skin contact.

2. Mitomycin C treated STO or SNL fibroblasts begin to die after they have been in culture for about 10 d. Fresh fibroblasts, therefore, have to be grown regularly, treated with mitomycin C, and either plated out immediately or frozen down for future use. PEF cells remain healthy for about 3 wk in culture after treatment. Dying feeders will not support ES cell growth. Unhealthy feeders look dark, grainy, and vacuolated. The edges of the cells become ragged and fine granular cell debris can be observed in the dish. Some cells detach and spaces appear in the monolayer. Healthy feeders are fibroblastic in morphology, i.e., spindle shaped, elongated, and have processes often growing from the poles of the cell. Healthy cells are colorless and surrounded by a glow around their borders. PEF cells remain spindle-shaped when mitomycin treated. STO/SNL cells may become more epithelial-like after treatment—a bit like crazy-paving. Change the

medium on the feeder cells every 4 d because they use up nutrients even though they are not dividing. Fibroblasts, particularly after mitomycin treatment, are very sensitive to trypsin. A few seconds (30–60 s) will be enough to bring them off the bottom of the dish. Quench the trypsin immediately with medium and rock the dish to mix this in or the cells are liable to be killed. To avoid this, trypsinize only a few dishes (3–4) at a time, rather than a larger number.

3. ES cells are confluent when the cells have covered most of the base of the dish and have made large colonies. Passage the plates before the cells begin to differentiate. Undifferentiated ES cells are very closely packed, with tight, virtually invisible cell–cell junctions. Cells that are differentiating either (1) take on a bumpy, epithelial like appearance, growing in a monolayer on the TC surface rather than in slightly raised colonies. Cell–cell junctions become loose and individual cells become clearly distinguishable. Alternatively, (2) groups of differentiating cells form high oval or spherical colonies that may be more loosely attached to the base of the dish. The central raised part of the colony often becomes covered with yellow grainy material that is made up of loose clumps of dead cells that float off. Distinct outlines around the colonies rapidly form discrete walls of cells that will later differentiate into giant trophoblast cells. These start to grow out from around the base of the colony into a flat plate-like structure, giving the effect of a "fried egg." Some colonies may float off the bottom of the dish, forming embryoid bodies.

4. The technical requirements for maintenance of a stable undifferentiated ES cell line are somewhat mysterious! There are no magic ES cell lines. Even the best ES cell populations can change karyotype/transmission properties in a dramatically short space of time. It is essential that they are maintained in a totipotent state, that they have a normal karyotype and, above all of course, that they give germline transmission under the culture conditions in your own laboratory.

5. If the ES cells on a plate have not grown to cover the plate evenly and there are just a few large colonies with a lot of empty space around them, the cells will need to be trypsinized, triturated, and replated on the same size feeder plate.

6. A confluent 90-mm plate of ES cells excluding feeders should yield approximately 1.3×10^7 cells.

7. A slow freezing container using isopropanol in the base (e.g., Nalgene "Mr. Frosty" from BDH) is recommended for the initial freezing of the cells at –70°C. This freezes the cells slowly to prevent ice crystal formation and cell membrane damage, improving cell viability during storage.

8. Additional tips for ES cell culture.
   • Feeder cells are larger than ES cells so can be recognized more easily under the microscope and excluded from counting.
   • Because of toxicity problems laboratory glassware must be absolutely clean: both low level detergent contamination and adsorbed heavy metals (e.g., chromium) are toxic to ES cells and so you can get "rogue pipets" that harm the cells. Disposable plastic pipets are the best choice, but are expensive.

- All the FCS used must be batch tested; many serum batches do not support ES cells growth very well and some are actively toxic.

9. The method by which transfected clones are to be screened for the occurrence of homologous recombination should be planned before a targeting vector is built. For replacement vectors with large flanking regions of homologous DNA, PCR screening becomes unfeasible and Southern blotting is both required and preferable because the likelihood of false positives is dramatically reduced. Southern blotting can be used to screen for homologous recombination in transfected clones when a probe external to one of the flanking regions of homology detects a restriction fragment on Southern blot which can be predicted to be altered after homologous recombination. It is preferable to have two external probes, one probe on each side of the targeting vector to be sure of correct homologous recombination. A third internal probe, such as the positive selection marker, is used to check that only one copy of the targeting vector is present, because randomly integrated additional copies can generate false phenotypes by disrupting other loci. The first round of screening should be done using one restriction enzyme and one probe. It is best to choose a "robust" restriction enzyme that can cut "dirty" DNA. The second round of screening putative positive clones should be performed with other restriction enzymes and probes. If the screening is done by PCR, a positive control vector has to be prepared before the targeting experiment and the PCR conditions have to be optimized. Such positive control vectors have to be handled very carefully because contamination could generate false-positive clones during the screening.

10. Slides are washed overnight in 5% glacial acetic acid in ethanol. Place in a more dilute acid-ethanol for long term storage. Wipe dry before use with a lint free cloth.

11. Normal murine cells have 40 acrocentric chromosomes. There should be no obviously abnormally sized chromosomes present (which may indicate the presence of translocations). We do not normally G-band our cells. For injection of clones into blastocysts, we expect each clone to have greater than 50% euploidy, but preferably 70–100%. All clones of less than 50% euploidy are discarded. Those with the highest percentage of euploidy are injected first.

12. Blastocyst collection is described elsewhere. Following injection, the blastocyst collapses, allowing the injected cells to be pressed tightly against the sticky inner cell mass cells within. In some cases, these injected recombinant cells will become integrated into the inner cell mass and continue growing to form part of the embryo.

13. As an early indication of the result of the procedure, variation can often be seen between the size of agouti pups and black pups in these litters, even before the coat color is evident. 129/Sv derived mice are generally slightly larger than C57BL/6 mice.

14. Recent application of the gene-targeting technique in the complement world has permitted the generation of mice deficient in specific complement components providing a deeper insight into the biological role of the complement system in vivo. Description of the various phenotypes observed in all the complement deficient mice is beyond the purposes of this chapter.

References

1. Lin, F. L., Sperle, K., and Sternberg, N. (1985) Recombination in mouse L cells between DNA introduced into cells and homologous chromosomal sequences. *Proc. Natl. Acad. Sci. USA* **82,** 1391–1395.
2. Smithies, O., Gregg, R. G., Boggs, S. S., Koralewski, M. A., and Kucherlapati, R. S. (1985) Insertion of DNA sequences into the human chromosomal beta-globin locus by homologous recombination. *Nature* **317,** 230–234.
3. Evans, M. J. and Kaufman, M. H. (1981) Establishment in culture of pluripotential cells from mouse embryos. *Nature* **292,** 154–156.
4. Martin, G. R. (1981) Isolation of a pluripotential cell line from early mouse embryos cultured in medium conditioned by teratocarcinoma stem cells. *Proc. Natl. Acad. Sci. USA* **78,** 7634–7638.
5. Thomas, K. R. and Capecchi, M. R. (1987) Site-directed mutagenesis by gene targeting in mouse embryo-derived stem cells. *Cell* **51,** 503–512.
6. Capecchi, M. R. (1989) Altering the genome by homologous recombination. *Science* **244,** 1288–1292.
7. Mansour, S. L., Thomas, K. R., and Capecchi, M. R. (1988) Disruption of the proto-oncogene int-2 in mouse embryo-derived stem cells: a general strategy for targeting mutations to non-selectable genes. *Nature* **336,** 348–352.
8. Hasty, P., Ramirez-Solis, R., Krumlauf, R., and Bradley, A. (1991) Introduction of a subtle mutation in the Hox-2.6 locus in embryonic stem cells. *Nature* **350,** 243–246
9. Valancius, V. and Smithies, O. (1991) Testing an "in-out" targeting procedure for making subtle genomic modifications in mouse embryonic stem cells. *Mol. Cell Biol.* **11,** 1402–1408.
10. te Riele, H., Maandag, E. R., and Berns, A. (1992) Highly efficient gene targeting in embryonic stem cells through homologous recombination with isogenic DNA constructs. *Proc. Natl. Acad. Sci. USA* **89,** 5128–5132.
11. Hasty, P., Rivera Perez, J., and Bradley, A. (1991) The length of homology required for gene targeting in embryonic stem cells. *Mol. Cell Biol.* **11,** 5586–5591.
12. Lovell-Badge, R. H. (1987) Introduction of DNA into embryonic stem cells, in *Teratocarcinomas and Embryonic Stem Cells: A Practical Approach* (Robertson, E. J., ed.), IRL, New York, pp. 153–182.
13. Bradley A. Production and analysis of chimaeric mice, in *Teratocarcinomas and Embyonic Stem Cells: A Practical Approach* (Robertson, E. J., ed), IRL, New York, pp. 113–152.
14. Papaioannou, V. and Johnson, R. (1993) Production of chimaeras and genetically defined offspring, in *Gene Targeting: A Practical Approach* (Joyner, A. L., ed.), IRL, New York, pp. 107–146.

17

Inherited Complement Deficiencies in Animals

Stuart Linton

1. Introduction

Animal models have been widely used to investigate the role of the complement system in a host of infectious and inflammatory diseases. Over the last 30 yr, a number of naturally occurring genetic deficiencies of complement components have been described in different species of laboratory animals. Where genetic studies have been undertaken, all these defects appear to follow an autosomal recessive pattern of inheritance and a number of them have been characterized at the molecular level. This chapter is a review of the complement-deficient strains available, experience derived from the use of such animals in experimental models in vivo, and of how the results of such studies have helped to define the effects of complement deficiency at different stages of the complement pathway.

2. Classical Pathway Deficiencies

Other than the preliminary report of a Wistar rat colony containing C4-deficient individuals (1), the only well-characterized deficiencies in early classical pathway components have been those described in C2 and C4-deficient guinea pigs (2,3) Both colonies were healthy in a normal breeding environment. C2 deficient animals were otherwise normocomplementemic, however, C4-deficient guinea pigs had variable reductions in the levels of C1 and C2 (4). Studies on the molecular mechanisms responsible for these deficiencies are summarized in **Table 1**. Both strains showed deficiencies in their immune responses to T-cell dependent antigens, characterized by lower antibody titers and impaired isotype switching after secondary immunization. These deficiencies were partially corrected by an increase in antigen dose (more readily in C2 deficient) (7). Of interest, some animals from both strains also showed serological evidence of immune complex disease such as raised total IgM levels and antiguinea

From: *Methods in Molecular Biology, vol. 150: Complement Methods and Protocols*
Edited by: B. P. Morgan © Humana Press Inc., Totowa, NJ

Table 1

Summary of Molecular Studies in Complement-Deficient Guinea Pigs

C2 deficient	C4 deficient	C3 deficient
No C2 detected by immunodiffusion, immunoelectrophoresis or functional assays; anti-C2 antibodies raised in deficient animals; Heterozygotes have 30–70% haemolytic activity	No C4 detected by immunodiffusion; anti-C4 antibodies raised in deficient animals. Heterozygotes have intermediate C4 levels by hemolysis and immunoprecipitation	5–10% C3 by ELISA. No anti-C3 antibodies raised in C3d. Heterozygotes have intermediate C3 levels; two types of C3 isolated from C3 deficient serum (only one type incorporated methylamine - a function of an intact thiolester bond)
Peritoneal macrophage culture No C2 haemolytic activity in supernatant from C2-deficient cells up to 72 h; Intermediate levels in heterozygotes (compared to normal)	**Peritoneal macrophage culture** No C4 haemolytic activity in homozygotes; 30–42% levels of normal in heterozygotes.	**Peritoneal macrophage culture** Similar levels of C3 from C3 deficient and normal cells; protein identical by SDS-PAGE.
	C4 deficient macrophage/HeLa fusion Some clones produced human C4 (possibly related to presence of 45 kD protein in macrophages which induced synthesis).	**Xenopus oocyte mRNA translation system** Same amounts of pro-C3/C3 from normal and C3 deficient mRNA; normal rate of secretion (two extra polypeptides detected in normal-70 and 48 kDa); C3 from C3 deficient was markedly more susceptible to trypsin proteolysis and showed reduced methylamine incorporation.
	Cell-free translation system Conflicting results -polysome-bound peptides immunoprecipitated by C4 in one study but not another; *neither* found C4/ pro-C4.	

230

Immunoprecipitation—In vitro cultures from four homozygotes: 3/4 - no C2; in 1 a mixture of LMW fractions identified (not known breakdown products); concentrated cell lysate preparation on SDS-PAGE yielded "C2" protein on SDS-PAGE with faster mobility than C2 from cultures of normal guinea pig cells

Conclusion

Possible structural abnormality in C2 (reviewed in **refs. 5** and **6**)

Northern blotting (liver)—5 kb Fragment in normal vs 7 kb in C4 deficient. **Southern blotting**—BamH1, 5.6 and 18.5 kb species from normal versus only 18.5 kb from C4 deficient.

Conclusion

Possible post-transcriptional processing defect of C4 (reviewed in **refs. 5** and **6**)

Northern and Southern blotting

No difference in C3 deficient animals.

DNA sequencing

Three nucleotide changes in C3 deficient (but all silent)

Conclusion

Possible disorder in primary/ higher order C3 structure (reviewed in **ref. 10**)

pig "rheumatoid factor," further supporting a role for the early classical pathway in the correct processing and solubilization of immune complexes (8).

C4 deficient guinea pigs showed greater susceptibility to the lethal effects of lipopolysaccharide (LPS), but were markedly more resistant to Forssman shock than complement-sufficient controls. In general, C4 deficiency seemed to have no effect on models of contact hypersensitivity, delayed hypersensitivity or passive cutaneous anaphylaxis (reviewed in **ref. 6**). Renal disease induced by injection of rabbit renal tubular basement membrane was identical in C4 deficient and normal animals, suggesting a lack of classical pathway contribution to this model of renal injury. Although in vitro, C4 deficient guinea pig serum is a less efficient bactericidal agent, no significant differences in clinical outcome were seen in vivo after challenge with a number of bacteria, other than a greater susceptibility to infection with *Treponema pallidum* (9).

3. C3 Deficiency

C3 occupies a central position in the complement pathway, mediating such diverse functions as convertase activity, opsonization, and anaphylatoxin production. C3 deficiency has been described in a number of laboratory animals (reviewed in **ref. 10**).

1. A possible C3 deficiency in guinea pigs was first described in 1919. These animals had a total lack of serum hemolytic activity, corrected by heated (but not cobra venom factor-treated) serum, and a markedly increased susceptibility to bacterial infection. These features are consistent with deficiency of C3 although "classical C3" at this time consisted of components C3, C5, C6, C7, C8, and C9. Unfortunately, this strain was lost through infection in the 1930s. More recently, a colony of partially C3-deficient guinea pigs has been described that have C3 levels of 5–10% of normal (*see* **Table 1**). These animals appeared clinically healthy despite reduced in vitro bactericidal activity against bacteria such as *Escherichia coli*. This strain of guinea pigs shared the abnormal antibody responses to T-cell dependent antigens seen in C2 and C4 deficient animals (*see* above), but showed none of the immunological abnormalities seen in these other two strains (reviewed in **ref. 10**). In a clinical model of immune-complex-mediated glomerulonephritis, C3-deficient guinea pigs showed subtle differences in histological patterns of renal deposit compared to normals, but these were solely related to lower antibody responses (11). A single study of cardiac xenotransplantation of rat organs into C3-deficient recipients showed increased graft survival (125 vs 67 d) beyond that observed in normal guinea pigs but further studies are lacking (12).

2. A colony of C3-deficient dogs has been described more recently (13). These animals lacked detectable C3 by enzyme-linked immunosorbent assay (ELISA), although they did display "C3-like hemolytic activity" at levels of 3–10% of normal. Although this activity eluted in similar fractions on column chromatog-

raphy to the C3 from normal animals, it lacked C3 immunoreactivity (reviewed in **ref.** *10*). The defect in C3 production has recently been shown to be because of a single base deletion resulting in frameshift and the generation of a stop codon with significant truncation of the normal C3 α-chain *(14)*. In contrast to C3-deficient guinea pigs, these animals suffered severe bacterial infections (up to 33% incidence at 6 yr) and major renal pathology. Of 20 dogs studied, 20% developed chronic renal failure, and all but one had histological evidence of type 1 membranoproliferative glomerulonephritis *(15)*. It may be that the small amounts of normal C3 in the guinea pig strain protected them from the full effects of C3 deficiency. C3 deficient dogs shared the aforementioned abnormalities in antibody responses. Other clinical studies have shown differences in these deficient animals in their responses to endotoxin challenge (with *greater* apparent organ dysfunction in C3 deficient) and prolonged cardiopulmonary bypass (with a *reduction* in the activation and pulmonary recruitment of neutrophils). In neither study was mortality or end organ damage significantly worse, suggesting that C3 has, at most, a partial pro- or antiinflammatory role to play in such situations *(16,17)*.

3. A partial C3-deficiency has recently been reported in a colony of C8-deficient rabbits (see below). A quarter of the animals tested had C3 levels around 10% of normal, with normal patterns of C3 on sodium dodecyl sulfate-polyacrylamide gel electrophoresis (SDS-PAGE) and by immunoelectrophoresis. C3-deficient animals had a lower survival rate at 3 mo and lower bactericidal activity in serum than normal rabbits *(18)*.

4. Terminal Complement Component Deficiencies
4.1. C5-Deficient Mice

Initial descriptions of inbred strains of C5-deficient mice date from the late 1960s *(19,20)*. Up to 39% of common inbred murine strains (including A/HeJ, DBA/2J, and B10.D2/Sn) carry this deficiency and lack detectable C5. Initial studies of in vitro synthesis of C5 by cell populations derived from C5-deficient animals found evidence of the production of abnormal molecular weight species of C5 that failed to be secreted (reviewed in **ref.** *21*). Northern blotting studies revealed significantly lower levels of RNA transcription in C5-deficient mice (both nuclear and cytoplasmic). These studies also revealed the presence of the larger "nuclear" transcript in the cytoplasm of C5-deficient animals suggesting that C5 deficiency might be related to the production of an abnormal primary transcript *(22)*. Sequence analysis of C5 cDNA from a number of C5-deficient strains has more recently identified a 2-bp deletion in an exon at the 5' end of the sequence, which would result in a predicted mRNA encoding a truncated 24.3 kD primary translation product *(23)*.

C5 deficient mice remain clinically well in normal laboratory housing. This, and the widespread availability of mouse strains with C5 deficiency, has resulted in a large volume of experimental work in both inflammatory and infectious models of disease (summarized in **Tables 2** and **3**). Some of these studies are,

Table 2
Review of the Use of C5-deficient Mice
in Inflammatory Disease Models (*see* **ref.** *21*)

Model	Outcome
Thyroid disease	
autoimmune thyroiditis induced by i/p injection of thyroglobulin.	Complement deficiency per se did not alter incidence or histological severity (number of strains studied).
Renal disease	
glomerulonephritis induced by injection of nephrotoxic rabbit serum.	a) No relationship between C5 deficiency and severity of nephritis (number of strains studied). b) Greater proteinuria and mortality in C5 deficient (no gross histological differences).
tubulointerstitial disease induced by kidney antigen.	No difference in outcome in C5 deficient.
Neurological disease	
myasthenia gravis induced by passive transfer of human IgG: *(24)*.	No difference in outcome in C5 deficient.
myasthenia gravis induced by injection of AchR: *(25)*.	Significantly greater incidence rates (78% vs 5%) and mortality in normal vs C5 deficient; similar titers of antibody response.
Pulmonary model	
immune-complex-mediated pulmonary injury.	Greater oedema, hemorrhage and neutrophil infiltration in normal animals compared with C5-deficient animals.
Transplantation	
cardiac xenograft: *(26)*.	Significant increase in graft survival in C5-deficient recipients (days vs min); further enhanced in C5 deficient by addition of i/v doxorubicin.
skin allografts.	No difference in graft survival in C5-deficient recipients compared to normal mice.
Dermatological disease	
bullous pemphigoid (induced by passive transfer of rabbit antimurine BP180 antibody): *(27)*.	Significant reduction in incidence in C5-deficient (0% versus 100% in normal) despite same antibody levels; normal incidence of disease restored in C5-deficient by coadministration of IL8 or C5a with pathogenic antibody (anti-BP180).

(*continued*)

Table 2 *(continued)*

Multiple organ dysfunction

Induced by i/p zymosan: *(28,29)*.	Lesser severity and lower mortality in acute stages in C5 deficient (two studies); only one study showed significant differences in later course or end organ damage amongst survivors.

Pancreatitis

(a) induced *by choline-deficient diet* supplemented with ethionine.	No difference in outcome in C5 deficient.
(b) *ligation*-induced model.	C5-deficient animals showed lower (though still substantial) degree of oedema and leucocyte infiltration, though only at later time points than 8 h.

Endotoxin

(a) TNFα priming, followed by i/v LPS at 30 min: *(30)*.	Lesser hypotension, better intestinal perfusion and lower mortality at 2–5 h in C5 deficient (0% vs 55%); increasing priming dose of TNFα abolished these differences.
(b) Priming with Complete Freund's adjuvant i/p, followed by LPS 2 wk later: *(31)*.	Lesser rise in pulmonary and hepatic permeability in C5 deficient early in model (<30 min); lower rise in TNFα levels in C5 deficient.

Arthritis model

Type II collagen arthritis [*see (32)* for overview].	Majority of studies in C5 deficient suggest that the presence of C5 is necessary, but not sufficient for disease induction, though a minority of studies are somewhat contradictory.

however, weakened by the use of noncongenic control strains, important as most of the models employed are also influenced by other genetic loci. C5 deficiency did significantly reduce inflammatory cell recruitment in the lung (whether elicited by immune complexes or pathogens) and skin (with the complete abrogation of experimental bullous pemphigoid), presumably related to the loss of C5a anaphylatoxic activity. In other tissues, the loss of C5 had weaker effects on the local inflammatory response (e.g., in the kidneys or pancreas), perhaps because of the contribution of alternative local mediators of cell recruitment and tissue damage. In other models, it appeared more likely that the greater contributory effect of C5 deficiency is via the loss of membrane attack complex (MAC) formation (e.g., in autoimmune myasthenia gravis and possibly collagen arthritis). Studies of infectious models have

Table 3
Review of the Use of C5-Deficient Mice in Infectious Disease Models (*see ref. 21*)

Model	Outcome
Extracellular bacterial models	
Corynebacterium kutscheri, i/v	Greater mortality in C5 deficient, but only at intermediate doses.
Escherichia coli, "complement-sensitive" strain, i/v	C5-deficient animals had greater bacterial organ load and lower in vitro serum bactericidal activity.
Streptococcus pneumoniae	
in vitro	Modest reduction in phagocytosis in vitro with C5-deficient serum.
i/p	LD50 10-fold lower for C5-deficient animals.
Staphylococcus aureus	
s/c	Larger area of necrotic skin after primary challenge in C5-deficient animals.
i/p	Greater mortality in C5-deficient animals.
aerosol	Marked reduction in lung clearance rate in C5-deficient animals despite similar recruitment of macrophages and PMNs in lungs of both groups. Organisms finally cleared in all animals. F1 "partially"-complement-deficient hybrids had intermediate clearance rates.
Mycoplasma sps..Mycoplasma canis (in vitro)	No bactericidal activity in C5-deficient serum.
Mycoplasma pulmonis (in vivo)	More severe arthritis in C5-deficient strains (noncongenic).
Neisseria gonococcus	
in vitro	Slower phagocytosis by C5-deficient peritoneal macrophages, but equally efficient intracellular killing.
in vivo	Greater susceptibility to disseminated infection in C5 deficient.
Borrelia turicatae, i/p	No difference in disease course in C5 deficient (noncongenic).

Pseudomonas aeruginosa
corneal infection
intratracheal
aerosol: (*33*).

No difference in disease course in C5 deficient.

Lower pulmonary recruitment of PMNs in C5 deficient.

Variable changes in PMN recruitment in C5-deficient animals, but all C5-deficient strains showed defective lung clearance; hybrid backcrosses linked C5 deficiency to inefficient clearance.

Intracellular bacterial model

Listeria monocytogenes
i/v challenge

Some studies reported greater mortality in C5-deficient mice at different doses; others did not. Study on a variety of strains *has* linked susceptibility to C5 deficiency. (B10.D2 may carry other "protective" loci (e.g., Lr2 allele) which complicate results.)

Slightly higher bacterial counts in liver and spleen during early infection in C5 deficient non-immune animals, but no differences seen in immune animals. Splenic counts could be "normalized" following transfer of bone marrow but not plasma from normal mice (?defect because of macrophage-associated C5).

i/p challenge.

Similar peritoneal exudation and in vitro function of peritoneal macrophages from C5 deficient and normal animals (immune or nonimmune).

Viral infections

Sindbis virus

Impaired clearance and greater mortality in C5-deficient mice.

Influenza A

Impaired clearance and greater mortality in C5-deficient mice.

Cowpox virus (*34*).

Impaired clearance and more severe ulceration in C5-deficient mice.

(continued)

237

Table 3 (continued)

Parasitic infections

Trypanosoma sps.

T. cruzi	No major differences in course of infection in C5-deficient mice.
T. musculi	No major differences in course of infection in C5-deficient mice.
Toxoplasma gondii	Earlier and greater mortality in C5-*sufficient* mice.
Plasmodium berghei	No major differences in course of infection in C5-deficient mice.
Schistosoma mansoni	No major differences in primary infection in C5-deficient mice.

Fungal infections

Candida

in vitro	Less efficient phagocytosis with phagocytes from C5-deficient mice.
in vivo	Earlier and greater mortality in vivo in C5-deficient strains.
Cryptococcus neoformans	Greater susceptibility in C5-deficient strains (noncongenic).
Paracoccidioides brasiliensis	No major differences in course of infection in C5-deficient mice.
Saccharomyces cerevisiae: (35).	Greater organ load and mortality in C5-deficient mice.

238

produced variable results, which is presumably a reflection of the relative importance of the complement system in the clearance of the pathogen under study. Observed in vitro deficiencies in opsonization and phagocytosis may translate directly into major differences in susceptibility to some pathogens in C5-deficient animals (e.g., pneumococcus and Candida), whereas in other infections, the effects of loss of complement effector function may be minimal or offset by other genetic resistance factors (e.g., parasitic infections and Listeria).

4.2. C6-Deficient Animals

Complete C6 deficiency has been described in colonies of rabbits and PVG-c rats *(36,37)*. A review of molecular and clinical studies in these animals is presented in **Table 4**. It is apparent that the particular model of disease chosen for study has significantly influenced the outcome in such studies (e.g., in renal settings). Findings from transplantation and coronary occlusion studies more consistently suggest a significant role for C6 (and by implication the MAC) in the etiology of hyperacute graft rejection and reperfusion injury in ischemic myocardium.

A preliminary description of a strain of Peru–Coppock mice deficient in C6 also has been reported *(54)*. These mice have been derived from wild-type murine species with a high spontaneous incidence of genetic mutation. No further information is yet available on these animals.

4.3. C8-Deficient Animals

A colony of C8-deficient rabbits have been described *(55)*. Intact, functional C8 consists of three polypeptide chains (α, β, and γ) and the complement deficiency in these animals was restored by the addition of the α–γ subunit of C8 isolated from human plasma. SDS-PAGE of serum from these rabbits revealed a β-subunit but no α–γ chain under nonreducing conditions, and the C8 γ chain could not be detected under reduced conditions in C8-deficient animals *(55)*. A similar type of α–γ subunit deficiency has been reported in occasional cases of human C8 deficiency. No gross differences have been reported between the results of Northern and Southern blots from normal and deficient animals leading to speculation that deficiency was due to a translational defect *(56)*. C8 deficiency in rabbits was also associated with greater early mortality and dwarfism. Despite an observed bactericidal defect in serum from these animals, no specific infectious complications have been observed. Some of this colony also manifested the partial C3 deficiency described above, though the two traits were not directly linked.

Two wild-type derived murine strains (MSM and Mae) have been described with a hemolytic defect corrected by human C8. These mice appear to have a deficiency in the C8 β-chain as judged by the capacity of their serum to restore lytic function to α–γ deficient human serum *(57)*.

Table 4
Overview of Animal Models of C6 Deficiency

C6 deficient rabbits:	C6 deficient rats:
Molecular mechanism No detectable C6 via hemolysis assays, rocket immunoelectrophoresis or isoelectric focusing (**38**).	*Molecular mechanism* <0.1% C6 detectable by ELISA. No C6 by Western blotting; Northern and Southern blotting revealed normal size bands at reduced intensity ? transcriptional defect or mRNA instability (**45**).
Clinical associations No infections; increased whole blood clotting time in plastic or glass (reviewed in **ref. 36**).	*Clinical associations* No infections; C2 levels 10% of normal (**37,45**).
Experimental models [*see* (**36**) for review]. 1) *Endotoxin: sublethal,* similar hemodynamic and hematological response; *lethal,* greater mortality in C6 deficient 2) **Serum sickness:** *acute,* no differences in clinical renal disease or proteinuria (**39**). 3) **Coronary A. occlusion:** *30 min occlusion,* smaller infarct in C6 deficient; *2 h occlusion,* equal size infarct but greater reperfusion and fewer arrythmias in C6 deficient (**40,41**).	*Experimental models* 1) **Renal disease:** *Active Heymann nephritis,* similar disease in normal and C6 deficient; *Anti-Thy-1.1 nephritis,* reduced cellular infiltrate/ proteinuria in C6 deficient animals and similar differences in nephritis also noted when extrahepatic production provided sole source of C6 (e.g., PVG-liver transplanted into PVG+ recipient) but *no* disease noted in single PVG+ renal transplant kidney in PVG-recipient (**45–47**).

2) **Transplant:** *cardiac xenograft,* C6-deficient recipients had significant increase in graft survival from donor guinea pigs or hamsters (*48,49*); effect prolonged by infusion of antineutrophil adhesion agent or serine protease inhibitor (*50,51*); *cardiac allograft,* longer graft survival in C6-deficient recipients (both nonsensitized and presensitized) (*52*); *liver transplant,* (PVG+ to PVG-) - normal C6 level by day 14, declined to zero by day 28 (anti-C6 detected); (PVG- to PVG+) - C6 levels fell to 30–40% by 24 h; BMT, (PVG+ to PVG-)- C6 restored to 5% at day 60 (*53*).

4) **Transplant:** *skin allografts*, greater survival (permanent in 2/23 cases) in C6 deficient; *cardiac xenografts:* C6-deficient recipients had longer graft survival (*42*)

5) **Miscellaneous:** *autoimmune thyroiditis*, milder disease in C6 deficient; EAE, similar pathology but C6 deficient suffered earlier paralysis (*43*); *Anti-GBM nephritis*, delayed onset in C6 deficient, but greater severity in later stages of model (*44*).

5. Deficiencies of Complement Inhibitors
5.1. Factor H Deficiency

Recently, a novel pig strain has been described with an autosomal recessive deficiency of the complement regulator, factor H (58). Homozygotes, which had inhibitor levels of 3% of normal by ELISA, rapidly succumbed to renal failure caused by type II membranoproliferative glomerulonephritis with a median survival of only 37 d (58). Heterozygotes had half normal factor H levels and were apparently disease-free. Factor H acts as an inhibitor of the alternative pathway convertase and it appears that, below a certain threshold, loss of inhibitory function causes massive renal complement deposition and activation. This strain has not been disseminated more widely despite its obvious importance for complement research.

6. Miscellaneous Deficiencies
6.1. C3a Receptor-Deficiency

Within the C4-deficient guinea pig colony mentioned previously, some animals have been shown to lack C3a receptor function, although the two deficiencies are not directly associated. These animals were initially characterized by a lack of in vitro responsiveness of certain derived cell populations (such as platelets, mast cells, and macrophages) to challenge with the anaphylatoxin, C3a (6). An in vivo study involving the infusion of C3a also revealed a difference in the early neutropenic response in these guinea pigs compared to normal animals. Although the cause of this deficiency is unclear, initial data has suggested that the C3a receptor may be present, but nonfunctional in such animals (59). The use of C3a receptor-deficient guinea pigs as cardiac donors in a xenotransplantation model failed to prolong graft survival in animals also receiving sCR1 (60).

7. Summary

Following the initial description of natural C5 deficiency in inbred mice, a growing number of complement component deficiencies in animals have been described, caused by a variety of genetic defects. Studies on such animals have contributed greatly to an understanding of the specific roles of classical, alternative and terminal pathway activation and inhibition in many infectious and inflammatory diseases. Further investigations in the more recently described models, in combination with the use of genetically engineered knockout mice (with targeted disruption of individual complement components and inhibitors), should continue to provide a fertile source of information regarding the role of complement in such experimental situations. This information is likely to have significant therapeutic implications for human disease.

References

1. Arroyave, C. M., Levy, R. M., and Johnson, J. S. (1977) Genetic deficiency of the fourth component of complement (C4) in Wistar rats. *Immunology* **33,** 453–459.
2. Bitter-Suermann, D., Hoffmamn, T., Burger R., and Hadding, U. (1981) Linkage of total deficiency of the second component (C2) of the complement system and of genetic C2-polymorphism to the major histocompatibility complex of the guinea pig. *J. Immunol.* **127,** 608–612.
3. Ellman, L., Green, I., and Frank, M. (1970) Genetically controlled total deficiency of the fourth component of complement in the guinea pig. *Science* **170,** 74–75.
4. Frank, M. M., May, J., Gaither, T., and Ellman, L. (1971) In vitro studies of complement function in sera of C4-deficient guinea pigs. *J. Exp. Med.* **134,** 176–187.
5. Colten, H. R. (1982) Biosynthesis of the MHC-linked complement proteins (C2, C4 and factor B) by mononuclear phagocytes. *Molec. Immunol.* **19,** 1279–1285.
6. Bitter-Suermann, D. and Burger, R. (1986) Guinea pigs deficient in C2, C4, C3 or the C3a receptor. *Progr. Aller.* **39,** 134–158.
7. Bottger, E. C., Hoffmann, T., Hadding, U., and Bitter-Suermann, D. (1985) Influence of genetically inherited complement deficiencies on humoral immune response in guinea pigs. *J. Immunol.* **135,** 4100–4107.
8. Bottger, E. C., Hoffmann, T., Hadding, U., and Bitter-Suermann, D. (1986) Guinea pigs with inherited deficiencies of complement components C2 or C4 have characteristics of immune complex disease. *J. Clin. Invest.* **78,** 689–695.
9. Wicher, K., Wicher, V., and Gruhn, R. F. (1985) Differences in susceptibility to infection with *Treponema pallidum* Nichols between five strains of guinea pigs. *Genitourin. Med.* **61,** 21–26.
10. Singer, L., Colten, H. R., and Wetsel, R. A. (1994) Complement C3 deficiency: Human, animal and experimental models. *Pathobiology* **62,** 14–28.
11. Schurman, S. J., McAdams, A. J., Beischel, L., Davis III, A. E., and Welch, T. R. (1995) C3-independent glomerulonephritis in guinea pigs: dependence upon primary humoral response. *Clin. Immunol. Immunopath.* **74,** 51–58.
12. Braidley, P. C., Dunning, J. J., Wallwork, J., and White, D. J. G. (1994) Prolongation of survival of rat heart xenografts in C3 deficient guinea pigs. *Trans. Proc.* **26,** 1259–1260.
13. Winkelstein, J. A., Cork, L. C., Griffin, D. E., Griffin, J. W., Adams, R. J., and Price, D. L. (1981) Genetically determined deficiency of the third component of complement in the dog. *Science* **212,** 1169–1170.
14. Ameratunga, R., Winkelstein, J. A., Brody, L., Binns, M., Cork, L. C., Colombani, P., and Valle, D. (1998) Molecular analysis of the third component of canine complement (C3) and identification of the mutation responsible for hereditary canine C3 deficiency. *J. Immunol.* **160,** 2824–2830.
15. Cork, L. C., Morris, R. M., Olson, J. L., Krakowsa, S., Swift, A. J., and Winkelstein, J. A. (1991) Membranoproliferative glomerulonephritis in dogs with a genetically determined deficiency of the third component of complement. *Clin. Immunol. Immunopath.* **60,** 455–470.

16. Quezado, Z. M. N., Hoffmann, W. D., Winkelstein J.A., Yatsiv, I., Koev, C. A., Cork, L. C., et al. (1994) The third component of complement protects against *Eschericia Coli* endotoxin-induced shock and multiple organ failure. *J. Exp. Med.* **179,** 569–578.

17. Gillinov, A. M., Redmond, J. M., Winkelstein, J. A., Zehr, K. J., Herskowitz, A., Baumgarter, W. A., and Cameron, D. E. (1994) Complement and neutrophil activation during cardioplumonary bypass: a study in the complement-deficient dog. *Ann. Thorac. Surg.* **57,** 345–352.

18. Komatsu, M., Yamamoto, K., Nakano, Y., Nakazawa, M., Ozawa, A., Mikami, H., et al. (1988) Hereditary C3 hypocomplementemia in the rabbit. *Immunology* **64,** 363–368.

19. Cinader, B., Dubiski, S., and Wardlaw, A. C. (1964) Distribution, inheritance and properties of an antigen, MuB1, and its relation to haemolytic complement. *J. Exp. Med.* **120,** 897–924.

20. Nilsson, U. R. and Müller-Eberhard, H. J. (1967) Deficiency of the fifth component of complement in mice with an inherited complement defect. *J. Exp. Med.* **125,** 1–16.

21. Rosenberg, L. T. and Tachibana, D. K. (1986) Mice deficient in C5. *Progr. Allergy* **39,** 169–191.

22. Wheat, W. H., Wetsel, R., Falus, A., Tack, B. F., and Strunk, R. C., The fifth component of complement (C5) in the mouse. *J. Exp. Med.* **165,** 1442–1447.

23. Wetsel, R. A., Fleischer, D. T., and Haviland, D. L. (1990) Deficiency of the murine fifth complement component (C5) J. Biol. Chem. **265,** 2435–2440.

24. Toyka, K. V., Drachman, D. B., Griffin, D. E., Pestronk, A., Winkelstein, J. A., Fischbeck, K. H., and Kao, I. (1977) Myasthenia gravis: Study of humoral immune mechanisms by passive transfer to mice. *N. E. J. M.* **296,** 125–131.

25. Christadoss, P. (1988) C5 gene influences the development of murine myasthenia gravis. *J. Immunol.* **140,** 2589–2592.

26. Sakakibara, N., Wolf, P., Kriett, J., Kapelanski, D. P., and Jamieson, S. W. (1996) Cardiac xenotransplantation into complement-5-deficient mice. *Trans. Proc.* **28,** 689–690.

27. Liu, Z., Giudice, G. J., Zhou, X., Swartz, S. J., Troy, J. L., Fairley, J. A., et al. (1997) A major role for neutrophils in experimental bullous pemphigoid. *J. Clin. Invest.* **100,** 1256–1263.

28. Nieuwenhuizen, G. A. P., Meyer, M. P. D., Hendriks, T., and Goris, R. G. A. (1995) Deficiency of complement factor C5 reduces early mortality but does not prevent organ damage in an animal model of multiple organ dysfunction syndrome. *Crit. Care Med.* **23,** 1686–1693.

29. Miller, C. G., Cook, D. N., and Kotwal, G. J. (1997) Two chemotactic factors, C5a and MIP-1α, dramatically alter the mortality from zymosan-induced multiple organ dysfunction syndrome (MODS): C5a contributes to MODS while MIP-1α has a protective role. *Molec. Immunol.* **33,** 1135–1137.

30. Hsueh, W., Sun, X., Rioja, L. N., and Gonzalez-Crussi, F. (1990) The role of the complement system in shock and tissue injury induced by tumour necrosis factor and endotoxin. *Immunology.* **70,** 309–314.

31. Barton, P. A. and Warren, J. S. (1993) Complement component C5 modulates the systemic tumor necrosis factor response in murine endotoxic shock. *Infect. Immun.* **61,** 1474–1481.

32. Mori, L. and de Libero, G. (1998) Genetic control of susceptibility to collagen-induced arthritis in T-cell receptor β-chain transgenic mice. *Arthr. Rheum.* **41,** 256–262.

33. Cerquetti, M. C., Sordelli, D. O., Bellanti, J. A., and Hooke, A. M. (1986) Lung defenses against Pseudomonas aeruginosa in C5-deficient mice with different genetic backgrounds.*Infect. Immun.* **52,** 853–857.

34. Miller, C. G., Justus, D. E., Jayaraman, S., and Kotwal, G. J. (1995) Severe and prolonged inflammatory response to localized cowpox virus infection in footpads of C5-deficient mice: Investigation of the role of host complement in poxvirus pathogenesis. *Cell. Immunol.* **162,** 326–332.

35. Byron, J. K., Clemons, K. V., McCusker, J. H., Davis, R. W., and Stevens, D. A. (1995) Pathogenicity of Saccharomyces cerevisiae in complement factor five-deficient mice. *Infect. Immunity* **63,** 478–485.

36. Rother, K., Rother, U., Muller-Eberhared, H. J., and Nilsson, U. R. (1966) Deficiency of the sixth component of complement in rabbits with an inherited complement defect. *J. Exp. Med.* **124,** 773.

37. Leenaerts, P. L., Stad, R. K., Hall, B. M., van Damme, B. J., Vanrenterghem, Y., and Daha, M. R. (1994) Hereditary C6 deficiency in a strain of PVG/c rats. *Clin. Exp. Immunol.* **97,** 478–482.

38. Goldman, M. B., Cohen, C., Stronski, K., Bangalore, S., and Goldman, J. N. (1982) Genetic control of C6 polymorphism and C6 deficiency in rabbits. *J. Immunol.* **128,** 43–48.

39. Parra, G., Takekoshi, Y., Striegel, J., Vernier, R. L., and Michael, A. F. (1992) Acute serum sickness in normal and C6 deficient rabbits: role of membrane attack complex. *Int. J. Exp.Path.* **73,** 299–312.

40. Ito, W., Schafer, H. J., Bhadki, S., Klask, R., Hansen, S., Schaarschmidt, S., et al. (1996) Influence of the terminal complement-complex on reperfusion injury, no-reflow and arrythmias: a comparison between C6-competent and C6-deficient rabbits. *Cardiovasc. Res.* **32,** 294–305.

41. Kilgore, K. S., Park, J. L., Tanhehco, E. J., Booth, E. A., Marks, R. M., and Lucchesi, B. R. (1998) Attenuation of interleukin-8 expression in C6-deficient rabbits after myocardial ischaemia/reperfusion. *J. Mol. Cell. Cardiol.* **30,** 75–85.

42. Chartrand, C., O'Regan, S., Robitaille, P., and Pinto-Blonde, M. (1979) Delayed rejection of cardiac xenografts in C6–deficient rabbits. *Immunol.* **38,** 245–248.

43. Inoue, K., Niesen, N., Biesecker, G., Milgrom, F., and Albini, B. (1993) Role of late complement components in experimental autoimmune thyroidits. *Clin. Immunol. Immunopath.* **66,** 1–10.

44. Groggel, G. C. and Terreros, D. A. (1990) Role of complement pathway in accelerated autologous anti-glomerular basement membrane nephritis. *Amer. J. Path.* **136,** 533–540.

45. van Dixhoorn, M. G. A., Timmerman, J. J., van Gijlswijk-Janssen, D. J., Muizert, Y., Verweij, C., Discipio, R. G., and Daha, M. R. (1997) Characterisation of complement C6 deficiency in a PVG/c rat strain. *Clin. Exp. Immunol.* **109**, 387–396.

46. Brandt, J., Pippin, J., Schulze, M., Hansch, G. M., Alpers, C. E., Johnson, R. J., et al. (1996) Role of the complement membrane attack complex (C5b-9) in mediating experimental mesangioproliferative glomerulonephritis. *Kidney Int.* **49**, 335–343.

47. Timmerman, J. J., Mieneke, G. A., van Dixhoorn, G. A., Schraa, E. O., van Gijlswijk-Janssen, D. J., Muizert, Y., van Es, L. A., and Daha, M. R. (1997) Extra-hepatic C6 is as effective as hepatic C6 in the generation of renal C5b-9 complexes. *Kidney Int.* **51**, 1788–1796.

48. Brauer, R. B., Baldwin III, W. M., Daha, M. R., Pruitt, S. K., and Sanfilippo, F. (1993) Use of C6-deficient rats to evaluate the mechanism of hyperacute rejection of discordant cardiac xenografts. *J. Immunol.* **151**, 7240–7248.

49. Lin, Y., Sobis, H., Vandeputte, M., and Waer, M. (1995) Natural killer cells, anti-body-dependent cellular cytotoxicity and complement synthesis by the xenograft itself play a role in xenograft rejection. *Trans. Proc.* **27**, 286–287.

50. Jakobs, F. M., Davis, E. A., Qian, Z. P., Liu, D. Y., Baldwin III, W. M., and Sanfilippo, F. (1997) The role of CD11b/CD18 mediated neutrophil adhesion in complement deficient xenograft recipients. *Clin. Transplantation* **11**, 516–521.

51. Jakobs, F. M., Davis, E. A., White, T., Sanfilippo, F., and Balswin, III, W. M. (1997) Prolonged discordant xenograft survival by inhibition of the intrinsic coagulation pathway in complement C6-deficient recipients. *J. Heart Lung Trans.* **17**, 306–311.

52. Brauer, R. B., Baldwin III, W. M., Ibrahim, S., and Sanfilippo, F. (1995) The contribution of terminal complement components to acute and hyperacute allograft rejection in the rat. *Transplantation* **59**, 288–293.

53. Brauer, R. B., Baldwin III, W. M., Wang, D., Horwitz, L. R., Hess, A. D., Klein, A. S., and Sanfilippo, F. (1994) Hepatic and extrahepatic biosynthesis of complement factor C6 in the rat. *J. Immunol.* **153**, 3168–3176.

54. Orren, A., Wallace, M. E., Hobart, M. J., and Lachmann, P. J. (1989) C6 polymorphism and C6 deficiency in site strains of the mutation-prone Peru-Coppock mice. *Complement Inflamm.* **6**, 295–296.

55. Komatsu, M., Yamamoto, K-I., Kawashima, T., and Migita, S. (1985) Genetic deficiency of the α–γ subunit of the eigth complement component in the rabbit. *J. Immunol.* **134**, 2607–2609.

56. Komatsu, M., Yamamoto, K-I., Mikami, H., and Sodetz, J. M. (1991) Genetic deficiency of complement component C8 in the rabbit: Evidence of a translational defect in expression of the α–γ subunit. *Biochem. Genet.* **29**, 271–274.

57. Tanaka, S., Suzuki, T., Sakaizumi, M., Harada, Y., Matsushima, Y., Miyashita, N., et al. (1991) Gene responsible for deficient activity of the β subunit of C8, the eigth component of complement, is located on mouse chromosome 4. *Immunogenetics* **33**, 18–23.

58. Hogasen, K., Jansen, J. H., Mollnes, T. E., Hovdenes, J., and Harboe, M. (1995) Hereditary porcine membranoproliferative glomerulonephritis type II is caused by Factor H deficiency. *J. Clin. Invest.* **95**, 1054–1061.

59. Gerardy-Schahn, R., Ambrosius, D., Saunders, D., Casaretto, M., Mittler, C., Karwarth, G., et al. (1989) Characterisation of C3a receptor-proteins on guinea pig platelets and human polymorphonuclear leukocytes. *Eur. J. Immunol.* **19,** 1095–1102.
60. Pruitt, S. K., Baldwin III, W. M., and Sanfilippo, F. (1996) The role of C3a and C5a in hyperacute rejection of guinea pig-to-rat cardiac xenografts. *Trans. Proc.* **28,** 596.

Appendices

1. Suppliers

Addresses/contact details	Products Relevant to the text
Accurate Chemical and Scientific Corp., 300 Shames Drive, Westbury, NY 11590 Tel. 516 433 4900 www.accuratechemical.com	Antibodies, density gradient media, standardised complement sera (rabbit, guinea pig etc.)
Advanced Research Technologies, 5230 Carroll Canyon Road, Suite 110, San Diego CA 92121 Tel. 619 552 8779 Fax. 619 552 8011	Anticomplement antisera; complement components and assays
Amersham PharmaciaBiotech, UK: Amersham Place, Little Chalfont, Bucks HP7 9NA, UK Tel. 0800 515313 USA: 2636 South Clearbrook Drive, Arlington Heights, IL 60005 Tel. 708 593 6300 www.amersham.co.uk	FPLC and other chromatography systems, columns, Chromatography matrices; Radiochemicals, etc.
American Type Culture Collection (ATCC), 12031 Parklawn Drive, Rockville, MD 20852-1776 Tel. 3018812600 www.atcc.org	Cell Lines, cDNA probes, etc.

From: *Methods in Molecular Biology, vol. 150: Complement Methods and Protocols*
Edited by: B. P. Morgan © Humana Press Inc., Totowa, NJ

AmiconLtd.   Ultrafiltration systems and
Europe: Upper Mill, Stonehouse,   membranes
Gloucestershire GL102BJ, UK
Tel. 01453 825181
USA: 72 Cherry Hill Drive, Beverly, MA 01915
Tel. 508 777 3622
www.amicon.com

Anachem Ltd.   Specialised filter membranes
Europe: 20 Charles Street, Luton, Bedfordshire
LU2 OEB, UK
Tel. 01582 745000
www.anachem.co.uk

Atlantic Antibodies Inc., 1990 Industrial   Immunochemicals
Boulevard, Stillwater, MN 55082
Tel. 612 439 9710
Fax. 612 779 7847
www.informagen.com

Becton Dickinson   Plastics, Flow cytometry
Europe: Between Towns Road, Cowley,   equipment and reagents,
Oxford OX4 3LY, UK   immunochemicals
Tel. 01865 748844
USA: 2350 Qume Drive, San Jose,
CA 95131-1807
Tel. 800 223 8226

Bachem Bioscience Inc.   Peptides, biochemicals and
Europe: Bachem (UK) Ltd, 69 High Street,   immunochemicals
Saffron Walden, Essex CB10 1AA, UK
Tel. 01799 526465
USA: Bachem Bioscience Inc,
3700 Horizon Drive,
King of Prussia, PA 19406
Tel. 610 239 0300
www.bachem.com

Behring Diagnostics (Dade Behring)   Anti-sheep erythrocyte antibody
UK: Walton Manor, Walton,   (amboceptor)
Milton Keynes MK7 7AJ, UK
Europe: Emil-von-Behring-Str. 76,
D-35041 Marburg, Germany
USA: 151 University Avenue,
Westwood, MA 02090
www.dadebehring.com

Biodesign Inc.
RR2, Box 1048
Kennebunkport, ME 04046
Tel. 207 967 4173
www.biodesign.com

Immunochemicals

The Binding Site Ltd.
Europe: P.0 Box 4073,
Birmingham B29 6AT, UK
Tel. 01214 142000
USA: Binding Site Inc., 5889 Oberlin Drive,
#101, San Diego, CA 92121
Tel. 619 453 9189
www.bindingsite.co.uk

Anti-C antisera; kits for
measurement of C components
and fragments (RID, ELISA,
etc.)

BioRad Laboratories Ltd.,
Europe: BioRad House, Maylands Avenue,
Hemel Hepstead HP2 7TD, UK
Tel. 01442 232552
USA: BioRad Labs, 2000 Alfred Nobel Drive,
Hercules, CA 94547
Tel. 1 800 424 6723
www.biorad.com

Electrophoresis Equipment,
molecular biology equipment;
Precast gels, Protein molecular
weight markers; secondary
antibodies (HRPO, AP etc.)

Boehringer Mannheim
Europe: Bell Lane, Lewes, East Sussex
BN7 1LG, UK
Tel. 01273 480444
USA: 5889 Oberlin Drive, #101;
San Diego, CA 92121
Tel. 619 453 9189
www.biochem.boehringer-mannheim.com

Antibody isotyping kits;
Biochemicals/immunochemicals

British Drug Houses (BDH)
Europe: Merck Ltd, Merck house,
Poole, Dorset, BH15 1TD, UK
Tel. 01202 664778
www.bdhinc.com

General laboratory equipment
and reagents

Calbiochem-Novabiochem (UK) Ltd.
Europe: Boulevard Industrial Estate
Padge Road, Beeston, Nottingham,
NG9 2JR, UK
Tel. 01159 430840
USA: PO Box 12087, San Diego, CA 92112-4180
Tel. 619 450 9600
www.calbiochem.com

Biochemicals, immunochemicals

Cymbus Bioscience Ltd.
Epsilon House, Chilworth Research Centre,
Southampton S01 7NS, UK
Tel. 01703 766280
www.cymbus.com

Immunochemicals

DAKO
Europe: Denmark House, Angel Drove,
Ely, Cambridge C117 4ET, UK
Tel. 01353 669911
USA: DAKO Corporation, 6392 Via Real,
Carpinteria, CA 93013
Tel. 805 566 6655
www.dakoltd.co.uk

Immunochemicals, secondary
antibodies, immunocytochemistry
reagents; reagents for flow
cytometry

Dynatech
Europe: Dynex Technologies,
Daux Road, Billingshurst, West Sussex
RH14 9SJ
Tel. 01784 251225
www.dynextechnologies.com

Flexible 96-well microtitre
plates, round or flat bottomed

European Collection of Animal Cell Cultures
(ECACC), PHLS Centre for Applied
Microbiology and Research, Porton Down,
Salisbury, Wilts SP4 0JG, UK
Tel. 01980 612512
www.camr.org.uk

Cell lines

Fisher Scientific
Europe: Bishop Meadow Road,
Loughborough, Leicestershire
LE11ORG, UK
Tel. 01509 231166
www.fishersci.com

General laboratory reagents and
equipment

Flowgen Laboratories
Europe: Lynn Lane, Shenstone, Lichfield,
Staffordshire WS14 OEE
Tel. 01543 483054
www.flowgen.co.uk

Ultrafiltration systems; filter
membranes (also suitable for
Amicon cells)

Gelman Sciences Ltd.
Europe: Brackmills Business Park,
Caswell Road, Northampton NN4 7EZ, UK
Tel. 01604 704704
www.pall.com/gelman/

Vacuum filtration systems,
Acrodise filters

Greiner Labortechnik Ltd.
Europe: Brunel Way, Stroudwater Business
Park, Stonehouse, Gloucestershire GL10 3SX.
Tel. 01453 825255
www.greiner-lab.com

Tissue culture plasticware and
other products

HybaidLtd
Europe: Action Court, Ashford Road, Ashford,
Middlesex TW15 1XB, UK
Tel. 01784 425000
www.hybaid.co.uk

Molecular biology equipment
and reagents

ICN Biochemicals
Europe: Unit 18, Thame Park Business Centre,
Wenman Rd, Thame, Oxfordshire OX9 3XA, UK
Tel. 0800 282474
USA: ICN Biomedical Research Products Division,
3300 Hyland Avenue, Costa Mesca, CA 92626
Tel. 800 854 0530
www.icnbiomed.com

Flexible ELISA plates

Invitrogen
Europe: Invitrogen BV, P.O. Box 2312,9704 CH
Groningen, The Netherlands
Tel. 00800 5345 5345
USA: Invitrogen Corporation,
1600 Faraday Avenue, Carlsbad, CA 92008
Tel. 1 800 955 6288
www.invitrogen.com

Molecular biology reagents and
systems; eukaryotic and
prokaryotic expression systems;
DNA libraries

Jackson Immunoresearch Labs.
Europe: Stratech Scientific Ltd,
61-63 Dudley Street,
Luton, Bedfordshire LU2 ONP, UK
Tel. 01582 481884
USA: Jackson Immunoresearch Labs,
872 West Baltimore Pike, P.O. Box 9,
West Grove, Pennsylvania 19390
Tel. 800 367 5296
www.jacksonimmuno.com

Antibodies and conjugates

Life Technologies Ltd.
Europe: 3 Fountain Drive,
Inchinnam Business Park,
Paisley PA3 9RF, Scotland, UK
Tel. 0800 269210
www.lifetech.com

Tissue culture plastics, media,
serum, etc.

Millipore
Europe: The Boulevard, Blackmoor Lane,
Watford, Hertfordshire, WD1 8YW, UK
Tel. 01923 816375
USA: Millipore Corporation, 80 Ashby Road,
Bredford, MA 01730 2271
Tel. 617 275 9200
www.millipore.com

Dialysis tubing, filter
membranes

Molecular Probes Inc.,
Europe: PoortGebouw, Rijnsburgweg 10,
2333 AA Leiden, The Netherlands
Tel. 3171523 3378
US: 4849 Pitchford Avenue, Eugene,
OR 97402-0469, UK
Tel. 541465 8300
www.probes.com

Fluorescent and other labelled
probes; calcein-AM, propidium
iodide etc; systems for quantify-
ing cell death, apoptosis, etc.

New England Biolabs,
Europe: Knowle Piece, Wibury Way, Hitchin,
Hertfordshire SG4 OTY, UK
Tel. 01462 420616
USA: 32 Tozer road, Beverly, MA 01915-5599
Tel. 978 927 5054
www.neb.com

Protein molecular weight
markers, molecular biology
reagents

Nunc
Europe: via Fisher Scientific (above)

Tissue culture plastics

Oxoid (Unipath Ltd)
Europe: Wade Road, Basingstoke,
Hampshire RG24 8PW, UK
Tel. 01256 816566
www.oxoid.co.uk

Biochemical Reagents; buffers
in tablet form—phosphate
buffered saline (PBS);
complement fixation diluent
(CFD) etc.

PAA Laboratories Ltd.
9 Derwentside Business Centre, Consett,
Co Durham DH86BP, UK
Tel. 01207 582993
USA: PAA Laboratories Inc, Marsh Creek
Corporate Center, Suite 328, 35 East
Uwchlan Avenue, Exton, PA 19341
Tel. 610 280 0474
www.paa-labs.com

Tissue culture reagents,
animal Sera

Pel-Freez Biologicals Inc.
PO Box 68, Rogers, AR 72757
Tel. 501636 4361
www.pel-freez-bio.com

Immunochemicals

Perkin-Elmer Ltd
Europe: Applied Biosystems Division,
Kelvin Close, Birchwood Science Park North,
Warrington, Cheshire, WA37PB
Tel. 01925 825650
www.perkin-elmer.com

Molecular biology equipment
and reagents, protein analysis
equipment and reagents

Pharmingen
Europe: via Becton Dickinson UK Ltd. (above).
Tel. 01865 748 844
USA: 11555 Sorrento Valley Road,
San Diego, CA 92121
Tel. 619 812 8800
www.pharmingen.com

Immunochemicals; specialists
in antibodies against nonhuman
targets

Pierce & Warriner
Europe: 44 Upper Northgate Street, Chester,
Cheshire, CHI 4EF
Tel. 01244 382525
USA: 3747 N. Meridan Road, P.O. Box 117,
Rockford, IL 61105
Tel. 815 968 0747
www.piercenet.com

Broad range of biochemicals;
numerous "kits" for protein
characterisation; protein assay
reagents; reagents for labelling
proteins (radiolabelling,
biotinylation, fluorescein
labelling, etc.).

Promega Ltd.
Europe: Delta House, Enterprise Road, Chilworth
Research Centre, Southampton S016 7NS, UK
Tel. 01703 760225
USA: Promega Corporation, 2800 Woods Hollow
Road, Madison, WI 53711 5399
Tel. 800 356 9526
www.promega.com

Molecular biology reagents

Quiagen Ltd.
Europe: Boundary Court, Gatwick Road,
Crawley, West Sussex RH10 2AX, UK
Tel. 01293 422911
USA: Labtrade Inc, 6157 NW 167th Street F-26,
Miami, FL 33015
Tel. 305 282 3818
www.quiagen.com

Molecular biology reagents

Quidel Inc.
Europe: PO Box 25, Betchworth,
Surrey R113 7YP, UK
Tel. 44 1737 844262
US: 10165 McKellar Court, San Diego, CA 92121
Tel. 619 552 1100
www.guidel.com

The most comprehensive range
of C-specific reagents; anti-C
antibodies, depleted sera, ELISA
kits for activation, etc

R&D Systems
Europe: 410 The Quadrant, Barton Lane,
Abingdon, Oxon OX14 3YS, UK
Tel. 01235 551100
USA: 614 McKinley Place NE,
Minneapolis, MN 55413
Tel. 612 379 2956
www.rndsystems.com

ELISA kits; cytokines and
cytokine antibodies, etc.

Sarstedt Ltd.
Europe: 68 Boston Road, Leicester LE4 1AW, UK
Tel. 01162 359023
USA: Sarstedt Inc., 1025 Saint James Church Road,
Newton, NC 28658-8627
Tel. 828 465 4000
www.sarstedt.co.uk

General laboratory equipment
and plasticware

Sartorius Ltd.
Europe: Longmead Business Centre,
Blenheim Road, Epsom, Surrey, KT19 9QN
Tel. 01372 737100
www.sartorms.com

Electrophoresis, nitrocellulose
membranes etc.

Schleicher and Schuell
Europe: Andermann and Co. Ltd.,
Laboratory Supplies Division, 145 London Road,
Kingston-Upon-Thames, Surrey, KT2 6NH
Tel. 01815 410035
www.s-und-s.de

Specialised membranes for
protein and DNA blotting.

Sera-Lab Ltd. (Harlan Sera-Lab)
Europe: Hophurst Lane, Crawley Down,
Sussex RH10 4FF,UK
Tel. 01342 716366
USA: via Accurate Chemical Co. (above)
www.labclinics.com/harlan.htm

Antibodies/sera

Serotec Ltd.
Europe: Serotec Ltd, 22 Bankside,
Station Approach, Kidlington, Oxford, UK
Tel. 01865 852700
USA: Serotec Inc., NCSU Centennial Campus,
Partners 1, 1017 Main Campus Drive, Suite 2450,
Raleigh, NC 27606
Tel. 1 800 265 7376
www.serotec.co.uk

Antibodies and conjugates,

Shandon
Europe: 93-96 Chadwick Road,
Astmoor Runcorn, Cheshire WA7 1PR, UK
Tel. 01928 566611
USA: 171 Industry Drive, Pittsburgh,
PA 15275-1034
Tel. 1 800 547 7429

Cytology/histology reagents and
equipment

Sigma Aldrich Co.
UK: Fancy Road, Poole, Dorset BH12 4QH, UK
Tel 01202 733114
US: PO Box 14508, St. Louis, MO 63178
Tel. 314 771 5750
www.sigma-aldrich.com

General laboratory reagents,
immunochemicals

Sorvall
Europe: Beckman Instruments (UK) Ltd, Oakley
Court, Kingsmead Business Park, London Road,
High Wycombe, Bucks HP1 1 1JU, UK
Tel. 01494 441181
USA: 2500 Harbour Boulevard, Box 3100,
Fullerton, CA 92634-3100
Tel. 1 800 742 2345
www.sorvall.com

Centrifuges

Stratagene
USA: 11011 North Torrey Pines Road,
La Jolla, CA 92037
Tel. 800 424 5444
www.stratagene.com

Molecular biology reagents

Surgipath
Europe: Venture Park, Stirling Way, Bretton,
Peterborough PE3 8YD, UK
Tel. 01733 333100
www.surgipath.com

Microscopy equipment and
reagents

Wako Pure Chemical Industries Ltd.   Immunochemicals, complement
1-2 Doshomachi, 3-Chome, Chuo-ku,   assay kits
Osaka 540-8605, Japan
Tel. 816 6203 3741.
USA: 1600 Bellwood Road, Richmond, VA 23237
Tel. 804 271 7677
www.wako-chem.co.jp

Whatman International Ltd   Chromatography columns and
Europe: St Leonards Road, Maidstone,   column matrices, Filter papers,
Kent ME16 OLS, UK   filters
Tel. 016222 682288
www.whatman.com

Zymed Laboratories Inc.   Immunochemicals
Europe: Cambridge Bioscience,
24-25 Signet Court, Newmarket Road,
Cambridge, C115 8LA, UK
Tel. 01223 316855
USA: 458 Carlton Court,
South San Francisco, CA 94080
Tel. 650 871 4494
www.zymed.com

2. Sources of C Components and Anti-C Antibodies

Component	Antibody sources
C1q	Protein: Ad, Q, S
	Polyclonal: Ac; Ad; BS; I; Q; S; Sl; St
	Monoclonal: Ac; Q
C1r	Protein: Ad, S
	Polyclonal: I; Q; S; Sl; St
	Monoclonal: nca
C1s	Protein: Ad, S
	Polyclonal: I; Q; S; Sl; St
	Monoclonal: nca
C4 (& fragments)	Protein: Ad, Q, S
	Polyclonal: Ac; Ad; BS; I; Q; S; Sl; St
	Monoclonal: Q; Pel
C2	Protein: Ad
	Polyclonal: BS; Ad; St; Q
	Monoclonal: nca
C3 (& fragments)	Protein: Ad, Q, S
	Polyclonal: Ac; Ad; BS; Dak; I; Q; S; Sl; St; others
	Monoclonal: Ac; Q; Pel

FB	Protein: Ad, Q, S
	Polyclonal: BS; Ad; Q; St
	Monoclonal: Q
FP	Protein: Ad, Q, S
	Polyclonal: Ad; Atl; BS; Q; S; SI; St;
	Monoclonal: Dak; Q
FD	Protein: Ad; Q; S
	Polyclonal: Ad; BS; Q; Wak;
	Monoclonal: nca
MBL	Nca
MASP1	Nca
MASP2	Nca
C5 (& fragments)	Protein: Ad, Q, S
	Polyclonal: Ac; Ad; BS; Dak; I; Q; SI; St;
	Monoclonal: Q
C6	Protein: Ad; Q; S
	Polyclonal: Ad; BS; Dak; I; Q; Sl; St; others
	Monoclonal: Q
C7	Protein: Ad, Q, S
	Polyclonal: Ad, BS; Dak; I; Q; Sl; St; others
	Monoclonal: Q
C8	Protein: Ad, Q, S
	Polyclonal: Ad, BS; Dak; I; Q; Sl; St; others
	Monoclonal: Ac; Q
C9	Protein: Ad, Q, S
	Polyclonal: Ad, BS; Dak; I; Q; Sl; St; others
	Monoclonal: Ac; Q
Cl inh	Protein: Ad, S
	Polyclonal: Ac; Ad; BS; Dak; I; Q; S; Sl; St; others
	Monoclonal: Ac;
FH	Protein: Ad, Q, S
	Polyclonal: Ad; BS; Q; St;
	Monoclonal: Q;
Fl	Protein: Ad, Q, S
	Polyclonal: Ad, BS; Q; St;
	Monoclonal: Q;
C4bp	Protein: S
	Polyclonal: Bid; BS; Q; St;
	Monoclonal: Bid; Q;
S-protein	Protein: S
(Vitronectin)	Polyclonal: BS; Q; St;
	Monoclonal: Cb; Q;
Clusterin	Protein: Q
	Polyclonal: nca
	Monoclonal: Q;

MCP (CD46)	Polyclonal: nca
	Monoclonal: Ag; Dak; St
DAF (CD55)	Polyclonal: nca
	Monoclonal: Ag; Cym; Dak; St
CR1 (CD35)	Polyclonal: nca
	Monoclonal: Ac; Ag; Cym; Dak; Sl; St
CD59	Polyclonal: nca
	Monoclonal: Ag; Dak; I; St
CR2 (CD21)	Polyclonal: nca
	Monoclonal: Ag; Dak; I; St
C3aR	Nca
C5aR (CD88)	Nca

Abbreviations: Ac, Accurate Scientific; Ad, Advanced Research Technologies; Ag, Autogen Bioclear Ltd.; AtI, Atlantic Antibodies Inc.; Bid, Biodesign inc.; BS, The Binding Site Ltd.; Cb, Calbiochem; Cym, Cymbus Bioscience Ltd.; Dak, DAKO Ltd.; I, ICN Ltd.; Pel, Pel-freez Biologicals Inc.; Q, Quidel Inc.; S, Sigma Ltd.; Sl, Sera-Lab Ltd.; St, Serotech Ltd.; Wak, Wako Chemicals Inc.; nca, not commercially available. Addresses listed in **Subheading 1**.

3. cDNA Accession Numbers for C Components, Regulators, and Receptors

Component	EMBL/Genbank accessions numbers
C1q	A chain, P02745
	B chain, P02746
	C chain, P02747
C1r	M14058
C1s	J04080
C4	K02403
C2	X04481
C3	K02765
fB	S67310
fP	X70872; X57748
fD	M84526
MBL	Y16581; Y16882
MASP1	D28593
MASP2	Y09926
C5	M57729
C6	X72177
C7	J03507
C8	α chain, M16974
	β chain, M16973
	γ chain, U08198
C9	X02176; K02766

C1 inh	NM_000062
fH	Y00716
fI	Y00318
C4bp	α chain, M31452
	β chain, L 11244
S-protein	X03168
Clusterin	X14723
MCP	Y0065 1; M58050
DAF	M31516
CR1	Y00816; X05309
CD59	X16447
CR2	Y00649; M26004
C3aR	U28488
C5aR	X57520; M62505

Index

From: *Methods in Molecular Biology, vol. 150, Complement Methods and Protocols*
Edited by: B. P. Morgan © Humana Press Inc., Totowa, NJ